Free Market Environmentalism

Free Market Environmentalism Revised Edition

Terry L. Anderson and Donald R. Leal

palgrave

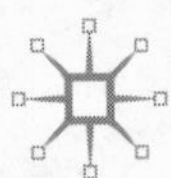

FREE MARKET ENVIRONMENTALISM

First published 2001 by
PALGRAVE™
175 Fifth Avenue, New York, N.Y. 10010 and
Houndmills, Basingstoke, Hampshire RG21 6XS.
Companies and representatives throughout the world

PALGRAVE is the new global publishing imprint of St. Martin 's Press LLC Scholarly and Reference Division and Palgrave Publishers Ltd (formerly Macmillan Press Ltd).

ISBN 0-312-23502-X hardback
ISBN 0-312-23503-8 paperback

Library of Congress Cataloging-in-Publication Data

Anderson, Terry Lee, 1946-
Free market environmentalism: revised edition / by Terry L. Anderson and Donald R. Leal.
p. cm.
Rev. ed. Of: Free market environmentalism.
Includes bibliographical references and index.
1. Environmental policy—United States. 2. Free enterprise—United States. 3. Natural resources United States—Management. I. Leal, Donald. II. Anderson, Terry Lee, 1946-

HC110.E5 .A66163 2000
333.7'0973—dc21 00-062696

A catalogue record for this book is available from the British Library.

Design by Westchester Book Composition

First edition: January 2001
10 9 8 7 6 5 4 3 2 1

Printed in the United States of America.

CONTENTS

To Bill Dunn, idea entrepreneur, friend, and fisherman

LIST OF TABLES AND FIGURES

Tables

Figures

ACKNOWLEDGMENTS

When we wrote the first edition of this book, the ideas of free market environmentalism were more theoretical than applied. Special thanks go to the hundreds of enviro-capitalists who have made free market environmentalism a pragmatic alternative to political environmentalism. To them we owe a debt of thanks for providing the examples that bring this book to life. We hope the ideas and examples will inspire other environmental entrepreneurs to move in this direction.

We owe a special thanks to professor Bruce Yandle, who is always teaching. In this case, he taught us to think about environmental problems as competing uses for resources, and this simple insight significantly changed the revision of the book.

Credit for taking the material from our word processors to the final form suitable for publishing goes to Michelle Johnson. Without her we would still be struggling to meet missed deadlines, and the reader would be struggling to make sense from our mistakes. She was assisted by others on the PERC staff who never get enough credit: Monica Lane Guenther, Dianna Rienhart, Colleen Lane, and Sheila Spain. We also want to thank Rachel Scarlett-Trotter and Mykel Matthews for their research assistance, and Rachel's mom, Lynn, for assisting our assistant.

Projects such as this require funding from people and foundations willing to make investments in ideas. Fortunately there is a long list of idea investors who made this project possible. We thank them in alphabetical order: Marty and Illie Anderson, Bill Dunn, the Earhart Foundation, Richard Larry of the Sarah Scaife Foundation, Thirty Five Twenty, Inc.*, and the Alex C. Walker Educational and Charitable Foundation. We also thank the Pacific Research Institute for its financial support, and Sally Pipes and Lisa Mac Lellan for their encouragement.

Finally, we thank Janet and Sandy for providing the most important support of all, their love.

CHAPTER 1

VISIONS OF THE ENVIRONMENT

For most people, markets are the cause of environmental problems, not the solution. The very notion of free market environmentalism is an oxymoron. Even conservative thinkers who support free enterprise and free trade find themselves uncomfortable with the idea of letting unfettered markets determine how and when natural resources are used. Markets may work fine to produce shoes or software, but the environment is somehow different and is too precious to be thrown to bulls and bears on Wall Street.

The feeling that markets and the environment do not mix is buttressed by the perception that resource exploitation and environmental degradation are inextricably linked to economic growth. This view, which first emerged with industrialization, builds on fears that we are running out of resources. After all, if resources are finite and production to meet material needs uses up some of those finite resources, the world's natural endowment must be getting more scarce. During the Industrial Revolution in England, the Reverend Thomas Malthus articulated this view, hypothesizing that exponential population growth would eventually overwhelm productivity growth and result in famine and pestilence. At the heart of Malthus' logic: population and growing consumption must eventually run into the wall of finite natural resources.

Earth Day 1970 started a plethora of doomsday predictions. Ecologist Kenneth Watt declared: "We have about five more years at the outside to do something," and Harvard biologist George Wald predicted that "civilization will end within 15 or 30 years unless immediate action is taken against problems facing mankind." Barry Commoner wrote: "We are in an environmental crisis which threatens the survival of this nation, and of the world as a suitable place of human habitation."[1]

Modern-day Malthusians have given such dire predictions an aura of credibility by using complex computer models to predict precisely when Malthusian calamities will occur. In the early 1970s, a group of scientists from the prestigious Massachusetts Institute of Technology concluded:

> If the present growth trends in world population, industrialization, pollution, food production, and resource depletion continue unchanged, the limits to growth on this planet will be reached sometime within the next one hundred years. The most probable result will be a rather sudden and uncontrollable decline in both population and industrial capacity.[2]

In a graph generated by their computer model, the scientific team showed that the "uncontrollable decline" would begin shortly after the turn of the century—in 2005, to be exact—with a precipitous decline in industrial output, food supplies, and population.[3]

The *Global 2000* report commissioned by President Jimmy Carter arrived at similar conclusions in 1980. It predicted:

> If present trends continue, . . . the world in 2000 will be more crowded, more polluted, less stable ecologically, and more vulnerable to disruption than the world we live in now. Serious stresses involving population, resources, and environment are clearly visible ahead.[4]

In every resource category, *Global 2000* predicted overuse and declines in quantity and quality.

Each year the Worldwatch Institute, headed by Lester R. Brown, makes headlines with its Malthusian predictions in its annual flagship publication, *State of the World*. In *State of the World 1998,* Brown states:

> Global land and water resources are not sufficient to satisfy the growing grain needs in China if it continues along the current development path. Nor will the oil resources be available, simply because world oil production is not projected to rise much above current levels in the years ahead as some of the older fields are depleted, largely offsetting output from newly discovered fields.[5]

He goes on to predict food crisis due to land and water scarcities: "The world is moving into uncharted territory on the food front, facing a set of problems on a scale that dwarfs those in the past."[6]

And year after year, the predictions from Worldwatch are proven wrong. The world does not run out of food; famine and pestilence are not on the increase; and the environment is generally not more, but rather less, polluted.[7] The one area where Worldwatch predictions tend to be more accurate is in the case of water, but even here they miss the mark. At the Second World Water Forum held in The Hague in March 2000, for example, the Food and Agriculture Organization of the United Nations (FAO) refuted claims that water shortages and food scarcity were imminent dangers. A recent FAO analysis of 93 developing countries revealed that the growing demands on global water are actually slowing. One FAO official commented that the "developing countries should be able to boost food production significantly by increasing irrigated cropland by about one third, yet using only 12 percent more water."[8]

The reason that Malthusian hypotheses are continually refuted is that they fail to take into account how human ingenuity stimulated by market forces finds ways to cope with natural resource constraints. As the late Julian Simon observed, the "ultimate resource" is the human mind, which has allowed us to avoid Malthusian cycles.[9] Human ingenuity is switched on by market prices that signal increasing scarcity and provide rewards for those who mitigate resource constraints by reducing consumption, finding substitutes, and improving productivity. For example, it was rising copper prices that helped stimulate the switch to fiber optics and satellite communication as substitutes for millions of miles of copper wire. Rising prices of aluminum and human ingenuity did more to reduce the thickness of drink cans and thus conserve resources used to produce and transport aluminum than did recycling regulations and campaigns. In fact, in the 1960s and the 1970s, the amount of aluminum in soft drink cans was reduced by 32 percent. These examples of aluminum and copper, two finite resources that Malthusians would doom to extinction, show that the human mind is a powerful tool, particularly when the incentives are right.

Julian Simon was so confident in the "ultimate resource" that he proposed a simple test to another modern-day Malthusian, Paul Ehrlich, author of *The Population Bomb.*[10] To test their competing theories, Simon offered a wager based on resource prices. He said that if Ehrlich's scarcity hypothesis was correct, it should be reflected in rising resource prices. On the other hand, if Simon's confidence in human ingenuity was correct, resource prices should fall. In 1980, Simon proposed that Ehrlich choose five resources and that they hypothetically purchase $200 worth of each. Ten years later they would see whether the $1,000 worth of resources had risen or fallen in value. If they had risen, Ehrlich would be vindicated and Simon would have to pay him the increase, or vice versa. Not surprisingly, Ehrlich accepted the wager and chose tungsten, copper, chrome, nickel, and tin. When they revisited the values in 1990, every single commodity chosen by Ehrlich had fallen in price even without adjusting for inflation. In fact the $1,000 bundle of five resources was worth less than $500 after adjusting for inflation. Ehrlich paid up, saying that he was still confident in his theory and that it would take longer for scarcity to manifest itself, but declined to accept Simon's offer to repeat the wager for $10,000.

Even for cases in which neo-Malthusians agree that human ingenuity can help mitigate the impacts of scarcity, they generally want top-down regulations and incentives from government to stimulate the ingenuity. For example, prices may encourage people to conserve energy or water, but energy and water markets cannot necessarily be depended upon to give the correct signals. In the case of ethanol at current and expected future prices, the corn-based alternative fuel would not be competitive with petroleum-based fuels, so the federal government subsidizes ethanol production. In the case of water, Americans would not be installing low-flow toilets were it not for federal regulations. And where prices are advocated as a stimulus to conservation, they are to be set by government, not by markets. According to Sandra Postel of the World Watch Institute,

> Wherever pricing and marketing fail to take into account the full social, environmental, and intergenerational costs of water use, some additional correction is needed. . . . In the case of fossil aquifers, such as the Ogallala in the U.S. High Plains or the deep desert aquifers in Saudi Arabia and Libya, a "depletion tax" might be levied on all groundwater extractions.[11]

From land to water to air, governmental control—which means political control—is seen as a necessary check on the environmental ravages of free markets.

This book will challenge the perception that free market environmentalism is an oxymoron and indeed argue that if we are to continue improving environmental quality in the twenty-first century, we must harness market forces. We offer an alternative way of thinking about natural resource and environmental concerns that contrasts information and incentives generated by markets and politics. In general, free market environmentalism emphasizes the positive incentives associated with prices, profits, and entrepreneurship, as opposed to political environmentalism, which emphasizes negative incentives associated with regulation and taxes.

At the heart of free market environmentalism is a system of well-specified property rights to natural and environmental resources. Whether these rights are held by individuals, corporations, nonprofit environmental groups, or communal groups, a discipline is imposed on resource users because the wealth of the property owner is at stake if bad decisions are made. Moreover, if private owners can sell their rights to use resources, the owners must not only consider their own values, they must also consider what others are willing to pay. In the market setting, it is the potential for gains from trade that encourages cooperation. Both the discipline of private ownership and the potential for gains from trade stand in sharp contrast to the political setting. When resources are controlled politically, the costs of misuse are more diffused and the potential for cooperation is minimized because the rights are essentially up for grabs.

Perhaps the best example of how these characteristics of private ownership can enhance environmental quality comes from the Nature Conservancy, the largest environmental group in the United States that depends on private landownership. When the conservancy obtains title to a parcel of land, it uses a formal system for evaluating whether the property has significant ecological value. Consider the case in which the Wisconsin Nature Conservancy was given title to beachfront property on St. Croix, Virgin Islands. One might think that the Nature Conservancy would go to great lengths to prevent development of oceanside property in the Caribbean. But, indeed, it actually traded the property for a much larger tract in Wisconsin and allowed selective beachfront development to occur under some protective covenants.

Why would an environmental group let this happen? The answer, in a word, is tradeoffs. As owner of the beach, the Wisconsin Nature Conservancy had to ask what would be gained and what would be sacrificed if development was prevented. The gain was obviously beachfront protection. The sacrifice may not be obvious to the casual observer, but it was obvious to the organization. At the time, the Wisconsin Nature Conservancy was actively trying to complete protection of a watershed in southern Wisconsin. It did not have the money to buy the last parcel of land needed to complete the protection, but it saw an opportunity to trade St. Croix beachfront for that rocky hillside. As owner of the St. Croix property, the Nature Conservancy faced the cost of just saying no or reaping the benefit of an entrepreneurial trade. The discipline and the incentives of private ownership forced the conservancy to make careful decisions and allowed it to accomplish its goal of saving a watershed. As a result, the Nature Conservancy's wealth in the form of environmental amenities was enhanced.

The emphasis of free market environmentalism on private ownership and decentralized decision making should not be taken to mean that there is no role for government. On the contrary, government has an integral role to play in the definition and enforcement of property rights. In the absence of the rule of law, the incentives inherent in private ownership disappear and with them goes the potential for environmental stewardship. With clearly specified titles obtained from land recording systems, strict liability rules, and adjudication of disputed property rights in the courts, market processes can encourage owners to carefully weigh costs and benefits and to look to the future.

Free market environmentalism conflicts with traditional environmentalism in its visions regarding human nature, knowledge, and processes. A consideration of these visions helps explain why some people accept this way of thinking as an alternative to bureaucratic control and why others reject it as a contradiction in terms.

Human nature: Free market environmentalism views man as self-interested. This self-interest may be enlightened to the extent that people are capable of setting aside their own well-being for close relatives and friends or that they may be conditioned by moral principles. But beyond this, good intentions will not suffice to produce good results. Developing an environmental ethic may be desirable, but it is unlikely to change basic human nature. Instead of intentions, good resource stewardship depends on how well social institutions harness self-interest through individual incentives.

Knowledge: In addition to incentives, good resource stewardship depends on the information available to individuals who make decisions about resource use. Free market environmentalism views this information as being time- and place-specific rather than general and concentrated in the hands of experts. Whether we are considering interactions among humans or interactions between humans and nature, the information necessary for good management varies significantly from time to time and from place to place. Certainly there is some knowledge that is general and concentrated in the minds of experts, such as the laws of physics or principles of ecology, but the complexity of ecosystems makes them impossible to model and therefore to manage from afar. Given that this type of time- and place-specific information cannot be gathered in a single mind or groups of minds that can account for the multiple interconnections of ecosystems, decentralized management guided by the incentives of private ownership becomes an alternative to centralized political control.

Process or solutions: These visions of human nature and knowledge combine to make free market environmentalism a study of process rather than a prescription for solutions. If man can rise above self-interest and if knowledge can be concentrated, then the possibility for solutions through political control is feasible. But if there are self-interested individuals with diffuse knowledge, then processes must generate a multitude of solutions conditioned by the costs and benefits faced by individual decision makers. By linking wealth to good stewardship through private ownership, the market process generates many entrepreneurial experiments; and those that are successful will be copied, while those that are failures will not. The question is not whether the right solution will always be achieved, but whether good decisions are rewarded and bad ones penalized.

These three elements of free market environmentalism—self-interest, information, and process—also characterize the interaction of organisms in nature. Since Charles Darwin's revolutionary study of evolution, most scientific approaches have implicitly assumed that self-interest generally dominates behavior for higher as well as lower forms of life. Individual members of a species may act in altruistic ways and may cooperate with other species, but species survival depends on adjustments to changing parameters in ways that enhance the probability of individual and species survival. To assume that man is not self-interested or that he can rise above self-interest requires heroic assumptions about *Homo sapiens* vis-à-vis other species.

Ecology also emphasizes the importance of time- and place-specific information in nature. Because the parameters to which species respond vary considerably within ecosystems, each member of a species must respond to time- and place-specific characteristics with the knowledge that each possesses. These parameters can vary widely, so it is imperative for survival that responses utilize the diffuse knowledge. Of course, the higher the level of communication among members of a species, the easier it is to accumulate and concentrate time- and place-specific knowledge. But a giant leap of faith is necessary if humans are to be able to accumulate and assimilate the necessary knowledge to manage the economy or the environment. Evidence from Eastern Europe underscores the problems that can arise with centralized management of either the economy or the environment.[12]

Continuing the analogy between ecology and free market environmentalism is instructive for thinking about the implications for policy. When a niche in an ecosystem is left open, a species will profit from filling that niche and will set in motion a multitude of other adjustments. If an elk herd grows because there is abundant forage, there will be additional food for predator species such as bears and wolves. Their numbers will expand as they take advantage of this profit opportunity. Individual elk will suffer from predation as elk numbers will be controlled. Plant species will survive and other vertebrates such as beavers will be able to survive. This is a process that no central planner could replicate because there is no best solution for filling niches and because each species is reacting to time- and place-specific information.

Comparing free market environmentalism with ecosystems serves to emphasize how market processes can be compatible with good resource stewardship and environmental quality. As survival rewards species that successfully fill a niche, increased wealth rewards owners who efficiently manage their resources.

Profits link self-interest with good resource management by attracting entrepreneurs to open niches. If bad decisions are being made, then a niche will be open. Whether an entrepreneur sees the opportunity and acts on it will depend on his or her ability to assess unique information and act on that assessment. As with an ecosystem, however, the diffuse nature of this information makes it impossible for a central planner to determine which niches are open and how they should be filled. If the information or incentives are distorted because property rights are incomplete or because decision makers receive distorted information through political intervention, then the market process will not necessarily generate good stewardship.

Visions of what makes good environmental policy will change only if we realize that our current visions are not consistent with reality. We must ask ourselves whether well-intentioned individuals armed with sufficient information dominate the political decisions that affect natural resources and the environment. Forest policy analyst and environmentalist Randal O'Toole answered this question in the context of the U.S. Forest Service.

> While the environmental movement has changed more than the Forest Service, I would modestly guess that I have changed more than most environmental leaders. . . . In 1980, I blamed all the deficiencies in the markets on greed and big business and thought that government should correct these deficiencies with new laws, regulatory agencies, rational planning, and trade and production restrictions. When that didn't work, I continued to blame the failure on greed and big business.
>
> About 1980, someone suggested to me that maybe government didn't solve environmental or other social problems any better than markets. That idea seemed absurd. After all, this is a democracy, a government of the people, and what the people want they should be able to get. Any suggestion that government doesn't work was incomprehensible.
>
> But then I was immersed in the planning processes of one government agency for ten years (sort of like taking a Berlitz course in bureauspeaking). I learned that the decisions made by government officials often ignored the economic and other analyses done by planners. So much for rational planning. Their decisions also often went counter to important laws and regulations. So much for a democratic government.
>
> Yet I came to realize that the decisions were all predictable, based mainly on their effects on forest budgets. . . .
>
> I gradually developed a new view of the world that recognized the flaws of government as well as the flaws in markets. Reforms should solve problems by creating a system of checks and balances on both processes. . . . The key is to give decision makers the incentives to manage resources properly.[13]

This book provides a "Berlitz course" in free market environmentalism that challenges entrenched visions. Because free market environmentalism depends on clearly specified property rights and accompanying price signals, it works better for some resources than for others. Markets can allocate land, water, and energy better than they can water quality or the global atmosphere. As we shall see, even on the western frontier, free market environmentalism was at work as cattlemen and farmers developed property rights to land and water and avoided the tragedy of the commons. In contrast, since one-third of the nation was set aside in the late nineteenth century for federal management, special interest politics has resulted in fiscal and environmental mismanagement. But even in the case of public lands, the implications of free market environmentalism are being utilized to improve incentives for federal managers by allowing them to charge user fees and to keep those fees for reinvestment in improving those lands. Similarly, in the case of water allocation, markets are improving efficiency and increasing instream flows for fish, wildlife, and recreation.

If land and water allocation are easy problems for free market environmentalism, pollution concerns challenge the paradigm. Again, however, the focus on property rights, with the accompanying right to be free from trespass by pollutants, provides a way of thinking how polluters can be held accountable. If it is possible to identify who is releasing pollutants into the soil, water, or air and to determine what the impacts of those pollutants are, then broad regulations can be replaced with negotiations between those who are harmed and those who are causing the pollution. In this way, property rights allow those who want cleaner land, water, or air to charge those who want to use it for waste disposal and hence make polluters accountable for costs they create. As we shall see in later chapters, measuring and monitoring pollution is not always easy, but just as barbed wire lowered the cost of fencing the western frontier, technology can lower the costs of holding polluters accountable.

By confronting our entrenched visions about the interface between markets and the environment, we can move beyond the status quo of political regulation and harness market forces to improve environmental quality. Included in these forces are environmental entrepreneurs who can fill market niches that allocate natural resources more efficiently and supply environmental amenities that we demand. These forces also require voluntary exchanges between property owners that promote cooperation and compromise, in contrast to political regulations that tear at the social fabric. In short, free market environmentalism offers an alternative that channels the heightened environmental consciousness into some win-win solutions that can sustain economic growth, enhance environmental quality, and promote harmony. Now join us in considering the details of this alternative vision.

CHAPTER 2

RETHINKING THE WAY WE THINK

John Maynard Keynes, the British economist responsible for changing the way we think about the role of government spending and taxing to even out business cycles, aptly described the impact of ideas, foretelling his own legacy:

> Both when they are right and when they are wrong, the ideas of economists and political philosophers are more powerful than is commonly understood. Indeed the world is ruled by little else. Even practical men, who believe themselves exempt from intellectual influences, are usually the slaves of defunct economists. Madmen in authority, who hear voices in the air, are distilling their frenzy from academic scribblers of a few years back.[1]

In the area of environmental economics, it was another British economist, A. C. Pigou, whose scribblings left their mark on modern environmental policy.[2] Pigou argued that because not all costs are taken into account by private decision makers, political intervention is necessary to correct what he saw as failures of the market. Hence, in the case of a paper mill disposing of its wastes by dumping them into the air or water, the mill is imposing unwanted costs on the rest of society, costs for which the mill owners are not held accountable. With these costs unaccounted for by polluters, private decision makers will overuse the water and air for waste disposal, and people who want to use these resources for other purposes (e.g., swimming or breathing) will bear the costs.

To counter this market failure and maximize the value derived from natural resources, Pigou called for taxes or regulations on polluters imposed through a political process. In his words,

> No "invisible hand" can be relied on to produce a good arrangement of the whole from a combination of separate treatments of the parts. It is therefore necessary that an authority of wider reach should intervene to

> tackle the collective problems of beauty, of air and light, as those other collective problems of gas and water have been tackled.[3]

Pigou believed that this authority should be given "to the appropriate department of central Government to order them [the polluters] to take action."[4]

Following the teaching of Pigou, economists and policy analysts have approached natural resource and environmental policy with the presumption that markets are responsible for resource misallocation and environmental degradation and that political processes can correct these problems. They assume there is a socially efficient allocation of resources that will be reached when "the appropriate department of central Government" correctly accounts for all the costs and benefits. One economist even went so far as to outline the conditions that would take society to its "bliss point."[5]

The purpose of this book is to challenge this traditional way of thinking and to provide a more realistic way of thinking about natural resource and environmental policy, thinking based on markets and property rights. This alternative recognizes and emphasizes the importance of incentives and of the costs of coordinating human actions. Rather than assuming that people are always altruistic, it presumes that self-interest prevails and asks how that self-interest can be harnessed to produce environmental goods that people demand. It does not assume that the costs of obtaining information or coordinating activities are zero or that there is perfect competition among producers. To the contrary, free market environmentalism focuses on how the costs of coordinating human actions (transaction costs, as economists label them) limit our ability to attain human goals through political processes and how markets can help overcome these costs. In this chapter, we proceed by considering the importance of incentives and transaction costs and then comparing incentives and transaction costs in market and political processes. Finally, the chapter addresses some specific arguments against free market environmentalism.

We emphasize from the outset that this way of thinking assumes that the environment's only value derives from human perceptions. Under this anthropocentric conception, the environment itself has no intrinsic value. People cannot manage natural resources for the sake of animals, plants, or other organisms because there is no Dr. Doolittle to "talk to the animals" and find out what is best for them. As long as humans have the power to alter the environment, they will do so based on human values—the only values that are ascertainable.

INCENTIVES AND TRANSACTION COSTS

In rethinking natural resource and environmental policy, two facts must be recognized. First, we cannot ignore the important role of incentives in guiding human behavior. No matter how well intended resource managers are, incentives affect their behavior. Like it or not, individuals will undertake more of an activity if the benefits of that activity are increased or if the costs reduced. This holds as much for bureaucrats and politicians as it does for profit-maximizing owners of firms or for citizens. Everyone accepts that managers in the private sector would dump production wastes into a nearby stream if they did not have to pay for the cost of their actions. Too often, however, we fail to recognize that the same elements work in the political arena. If a politician is not personally accountable for allowing oil development on federal lands or for the environmental impact of building dams on naturally-flowing rivers, we can expect too much oil development or too many dams. Moreover, when the beneficiaries—call them special interest groups—of these policies do not have to pay the full cost, they will demand more of them from their political representatives.

Once incentive effects are recognized, we can no longer rely on good intentions to generate good resource stewardship and environmental quality. Even if the superintendent of a national park believes that grizzly bear habitat is more valuable than additional campsites, his good intentions will not necessarily result in the creation of more grizzly bear habitat. Hence, Grant Village, a tourist facility in the southern part of Yellowstone National Park, was built in the middle of prime grizzly bear habitat because politics, not science, dominated the decision.[6] In a political setting where commercial interests have more influence over a bureaucrat's budget, his peace and quiet, or his future promotions, good intentions by the bureaucrat will have to override political incentives if grizzly bear habitat is to prevail. Although possible in some cases, there is ample evidence that good intentions are not enough.

If a private resource owner believes that grizzly bear habitat is more valuable and can capture that value through a market transaction, then politics will not matter. Moreover, if those demanding the preservation of grizzly habitat are willing to pay more than those who demand campsites, then incentives and information reinforce each other.

Traditional thinking about natural resource and environmental policy has tended to emphasize incentive problems inherent in markets but to ignore them in the context of political processes. In markets profit maximizers are continually on the lookout for ways to externalize costs, but in the political process they are presumed to be on the lookout for the public interest. Consider the approach taken in one natural resource economics textbook that describes government as "a separate agent acting in the social interest when activity by individuals fails to bring about the social optimum. . . . we discuss some limits of this approach, but it permits us to abstract from the details of the political process."[7] To abstract from the details of the political process ignores incentives inherent in that process. Daniel Bromley claims that government agencies are

> politically responsible to the citizenry through the system of . . . elections and ministerial direction. However imperfect this may work, the *presumption* must be that the wishes of the full citizenry are more properly catered to than would be the case if all environmental protection were left to the ability to pay by a few members of society given to philanthropy.[8]

Why must we "presume" that the "wishes of the full citizenry are more properly catered to"? Moreover, what does "full citizenry" mean? Is there unanimous consent? Does a majority constitute the "full citizenry" when voting turnout is traditionally low? Bromley also charges that "claims for volitional exchange are supported by an appeal to a body of economic theory that is not made explicit," but there is little made explicit when we "abstract from the details of the political process" by presuming "that the wishes of the full citizenry are more properly catered to" in the political process. In short, the lens of free market environmentalism forces us to realize that incentives matter everywhere.

Second, free market environmentalism focuses our attention on the costs of coordinating human activities. The scribblings of Nobel laureate Ronald Coase brought the importance of transaction costs into the forefront of policy analysis.[9] Coase's important point was that transaction costs in the marketplace and in the political arena explain why individuals may not always be able to resolve their competing uses of resources and the environment. He explained that, in a world of zero transaction costs, markets would work perfectly because producers and consumers would know all. Producers would always supply consumers with what they want, and consumers would always be able to hold producers accountable for any costs created by production. In the political arena, zero transaction costs would also yield perfect results because citizens would have no problem communicating their demands to politicians or knowing whether their demands were being met.

Of course, transaction costs are not zero in the real world. Producers do not always know what consumers want. Consumers do not always get what they expect. And people who use resources or dispose of garbage in the environment are not always held accountable for their actions. It is this lack of accountability that explains almost all concerns about natural resource stewardship and environmental quality.

Consider what happens when two people engage in a trade where one offers meat raised on private land in exchange for fish caught in the open ocean. The supplier of meat must consider the impact of his grazing cattle on the future productivity of the land. If he grazes too many cattle this year, there will be less grass next year, and possibly no grass at all. The fish supplier, on the other hand, faces a very different set of costs. Catching fish this year means those fish will not have an opportunity to grow larger and to reproduce, but in the open ocean a fish left for tomorrow will be caught by another fisher. Hence, each fisher ignores the future value of the fishery and overharvests today. After all, a fish not taken will be caught by someone else. Indeed, taking a fish today imposes costs on all fishers tomorrow because fish will be smaller and will not be reproducing, but these costs are spread among all fishers, while the benefits redound to the individual.

This problem is known as the tragedy of the commons.[10] If access to a valuable resource such as an ocean fishery is unrestricted, people entering the commons to capture its value will ultimately destroy it. Even if each individual recognizes that open access leads to resource destruction, there is no incentive for him to refrain from harvesting the fish. If he does not take it, someone else will, and therein lies the tragedy.

A similar tragedy occurs where there is open access to water as a medium for waste disposal. When consumers contract with paper producers to supply paper, some waste is inevitable, and the producers will seek the cheapest way to dispose of the waste. If they choose to dump it into the water, waste disposal will compete with other uses such as swimming. As far as swimmers are concerned, the tragedy of the commons has resulted in too much use of water for waste disposal.

The tragedy of the commons can also result in not enough production of a good thing if people can free ride on the actions of others. For example, if one individual or group sets aside land for biodiversity, the benefits of that biodiversity may redound to many people who did nothing to help provide it. As long as third parties can enjoy environmental amenities without paying for them, there is the potential for a free ride and therefore the possibility that the amenities will be underproduced. In other words, if third-party costs result in too much pollution, third-party benefits (free riding) result in too little production of environmental amenities.

All of these tragedies raise two questions: who has what rights and what are the costs associated with defining and enforcing those rights? Where rights are clearly defined and easily enforced, as in the case of surface land, there is no tragedy because entry is limited by the owner's fence. If party A dumps his garbage on party B's land, party B can enforce his right against trespass. On the other hand, where rights are not well defined or easily enforced, as with the right to clean air, trespass is much more difficult to prevent. It is much more difficult to identify who owns the fish, the water, or the air than it is to specify who owns the land, making enforcement against trespass a much tougher task. If the value of preserving wilderness is derived mainly by those who wish to hike in that wilderness, then the landowner can install pay booths at entrances and collect payment for the services he is providing. If the value is derived mainly by people who enjoy sitting in their offices thinking about the existence of wilderness, however, it will be more difficult for the owner/provider of wilderness to collect for his efforts.

Reenter transaction costs. If it were costless to organize to restrict entry into the commons, the tragedy would never occur. It does occur, however, because organizing to restrict entry is costly. People will not always know the impact of their actions until it is too late and the commons is overexploited. Even if they do recognize the potential for tragedy, the costs of organizing can be high. Bargaining to agree on who will fish and when and forming binding agreements to restrict fishing will be costly especially to the extent that detection of violators is difficult. Because these two types of costs—information costs and contracting costs—are pervasive in both market and political solutions to environmental problems, it is important that we examine them in more detail.

Information costs are the costs of articulating or measuring the values that humans place on goods and services they desire from their limited resources and of knowing how one person's actions impinge on the values of another. These are the costs that each of us incurs when trying to decide whether we would be happier spending more of our budget on housing, health care, or outdoor recreation. They are the costs companies incur in trying to determine whether there is a market for their products. They are the costs of knowing whether barbecue smoke wafting from your neighbor's yard is sufficiently carcinogenic to warrant asking your neighbor to stop polluting. They are the costs that politicians incur when deciding whether constituents would be better off with more national defense, more public lands, or lower taxes.

If all of these costs were zero, solving environmental problems would be much easier, but they are not and people must find ways of articulating and discovering information. In families and other small groups of people who care about one another, for example, intimate knowledge of one another's values may suffice to allow group members to take into account the preferences of others. Beyond such groups, we rely on communication processes in which there is far less personal knowledge of the values of others. In markets, for example, consumers will have to discover the value of goods they wish to purchase by doing research on product quality, and producers will attempt to supply information about their goods through advertising. Ultimately consumers express their values by offering money for goods and services, and suppliers decide whether these offers are sufficient to cover their costs of supplying. In a democratic political arena, we must do our research on candidates and then communicate our values by voting, protesting, letter writing, and contributing to campaigns, and the politicians must decide whether they can or want to meet the competing demands for goods and services subject to political constraints.

Even if we solve the information problem, we contract with one another to achieve our goals. These contracting costs include bargaining on the terms of the agreement and enforcing the terms of the agreement. Consumers need to ensure delivery of the goods and services for which they pay. Was the price paid commensurate with the expected quality? Were the goods or services delivered on time? Did suppliers charge more than consumers ultimately realized the product was worth? Likewise, suppliers must ensure that they are paid for services rendered. Was the payment on time? Did it cover the costs? In the political arena, did the politician deliver on his or her campaign promises? Are the citizens paying their taxes and abiding by the rules and regulations established by government? All such contracting costs make it more difficult for consumers and suppliers—whether in markets or politics—to coordinate with each other to enjoy gains from trade.

MARKETS AND POLITICS COMPARED

Though there is a myriad of processes for coordinating human interaction in order to benefit from potential gains from trade (for example, families, clubs, or totalitarian states) and to prevent the tragedy of the commons, we compare and contrast transaction costs in the context of two—market processes and political processes. The important point is that although the information and contracting costs outlined above are endemic to both coordination processes, costs and benefits faced by decision makers differ systematically between the two and thus affect incentives and outcomes.

Information Costs

First, consider information costs. In a world of scarce resources, private and political resource managers must obtain information about the relative values of alternative uses of everything from land to wildlife to air. When one resource use rivals another, tradeoffs must be made, and resource managers can only make these tradeoffs based on the information they receive, or on their own personal values. For example, if timber managers believe lumber is more valuable than wildlife habitat, they will cut trees. Timber managers may know how fast trees grow under different soil and climate conditions, but they cannot know the value of that growth without incurring some cost of surveying how consumers value the wood.

In the marketplace, prices provide an objective measure of subjective preferences and are therefore an important source of information about subjective values. Because each of us places different value on environmental amenities, there must be some way of quantifying and aggregating those values. Some see a forest as a place for quiet hikes, while others see it as a place for snowmobiling. Some see a rain forest as a jungle that, when cleared, can grow crops, while others see it as a source of biodiversity. Psychology can tell us a little about how these values are formed and influenced by peers, parents, advertising, genetics, and so on, but ultimately they are subjective to each individual.

Once individuals undertake market trades to achieve their desires, their bids provide an objective measure of these subjective values because bidders must give up one thing of value to obtain another. In the case of timberlands, private and public timber managers can obtain relatively comprehensive information on the value of wood from a lumberyard, where people offer money, which could be used to purchase other goods and services, for the wood products they value more highly. In the absence of markets for wildlife habitat or hiking trails, however, obtaining values is more costly. Nonetheless, private timber managers in a company such as International Paper obtain information on the value of wildlife amenities through an active market for hunting, camping, and other recreation on their private lands. When leasing its land for these activities, the company faces a tradeoff between timber harvesting, which produces revenue from wood products, and recreational land uses, which produce revenue from not cutting trees. Decisions on land use are driven by the differences in potential profit between the two activities.

Prices also allow a measure of efficiency through profits and losses. If a shareholder wants to know how well the management of his firm is performing, he can at least consult the profit and loss statement. This may not be a perfect measure of performance, but continual losses suggest that actual results differ from desired results. This can indicate to the shareholder that he should consider alternate managers who can produce the product at a lower cost or that he should reconsider the market for the product. Unlike the political sector, where the output of government is not priced and where agency performance is not measured by the bottom line, profits and losses in the private sector provide concise information with which owners can measure the performance of their agents.

In the public sector, on the other hand, there are few market prices and no profits to motivate decisions. Loggers compete in auctions for timber sales and thus provide some objective measure of market values, but recreational users of public lands generally pay little or nothing for the services they receive. Hence information on recreational values must be revealed through the political process. Special interest groups may articulate their demands through voting, campaign contributions, and letter-writing campaigns, to mention a few. In this process, lumber companies might argue that timber harvesting is the most important use of public land, while environmental groups will argue that wilderness values should trump all other values, including logging.

Hence free market environmentalism identifies systematic differences in the way information about subjective values is communicated in markets and politics. In the marketplace, prices lower information costs by converting subjective values into objective measures. In a democratic political process, the main counterpart to prices for signaling values is voting. Voting is a signal that, at best, communicates the subjective values of the median voter and, especially given that voter turnout is often low and representative of organized interest groups, communicates the subject values of special interest groups. While information costs are positive in both processes, prices offer a low-cost mechanism for articulating subjective values, and connect the person paying the price with the actual cost of the product or service.

As a solution to the information problem, some policy analysts embraced the idea of scientific management. This idea first surfaced in the United States with the formation of the U.S. Forest Service in the late nineteenth century. Ostensibly, scientific management directed at the federal level was supposed to be the answer to the perceived exploitation of U.S. forests.[11] Because the main task of the Forest Service was to manage forests for future wood fiber production in accordance with the best silva cultural techniques, there was little need to consider other values. However, as citizens have begun to demand other products and services from political lands, professional foresters have been forced, by either politics or legislation, to consider other values and to trade off between multiple uses.

Making these tradeoffs, in the economist's framework, is a simple matter of comparing the additional value of one use to the additional value of another. The calculus is simple; if the additional value of shifting a resource from one use is greater than the value in the use from which it is being taken, then reallocation will be prudent. In other words, if the marginal benefits are greater than the marginal costs, do it.

In this analysis, there are many margins for adjustments and few decisions that have all-or-nothing consequences. Put simply, neither demand nor supply is insensitive to price changes. If prices rise, then consumers will adjust by shifting consumption to the nearest substitutes, and suppliers will adjust by shifting to other inputs or technologies, or by producing other products.

The logic of this analysis combined with models and computers capable of simulating resource use can lure policy analysts into thinking that efficient resource management is simple. Unfortunately, such logic and simplification are not helpful guides because they mask transaction costs and incentives.

Consider the case of multiple-use management of the national forests, where scientific managers are required to balance timber production, wildlife habitat, aesthetic values, water quality, recreation, and other uses to maximize the value of the forest. Scientific managers, not motivated by profits or self-interest, who are armed with the economic concept of marginal analysis, are assumed to be omniscient, analytical, and impartial.[12] But as F. A. Hayek pointed out many years ago, "the economic problem of society is . . . not merely a problem of how to allocate 'given resources' if 'given' is taken to mean given to a single mind which deliberately solves the problem set by these 'data.' "[13]

The problem is that obtaining the value data to make the necessary tradeoffs is no small task because of information costs. Scientific management assumes that values are known or can be discovered and, therefore, that there is also an efficient solution waiting to be discovered. Thomas Sowell describes this view of traditional resource economics as it relates to scientific management:

> Given that explicitly articulated knowledge is special and concentrated . . . the best conduct of social activities depends upon the special knowledge of the few being used to guide the actions of the many. . . . Along with this has often gone a vision of intellectuals as disinterested advisors . . .[14]

As analytical tools, economic models focus on the importance of marginal adjustments, but they cannot instruct managers in which tradeoffs to make or which values to place on a particular resource. In the absence of objective measures of subjective individual values, the marginal solutions derived by sophisticated efficiency-maximization models are unachievable ideals. Though these models add sophistication to decisions and give them an aura of authority and correctness, they cannot be effectively implemented in a world where political forces drive incentives.

No matter how rational or comprehensive the models may be, such models still necessitate obtaining costly information. Here again Hayek's insights are valuable, for he saw the allocation problem as one of "how to secure the best use of resources known to any of the members of society, for ends whose relative importance only these individuals know. Or, to put it briefly, it is a problem of utilizing knowledge not given to anyone in its totality."[15] As he well understood, subjective human values are best revealed through human action in accordance with those values. What form that action takes—for example, bidding or lobbying—will depend on incentives that in turn depend on the allocation system.

In contrast to scientific management, the market process generates information on the subjective values as individuals engage in voluntary trades. The decentralized decisions made in markets are crucial because "practically every individual has some advantage over all others in that he possesses unique information of which beneficial use might be made, but of which use can be made only if the decisions depending on it are left to him or are made with his active cooperation."[16] Once we understand that most knowledge is fragmented and dispersed, then we can understand that "systemic coordination among the many supersedes the special wisdom of the few."[17]

Traditional economic analysis has failed to recognize this fundamental point. The information necessary for "efficient" resource allocation depends on the knowledge of what Hayek called the special circumstance of time and place.[18]

Contracting Costs

Contracting costs also differ systematically between market and political processes. In a market transaction in which the Nature Conservancy purchases conservation easements to prevent land development, it must negotiate what land is involved, what uses are acceptable, and what price will be paid. At the same time the landowner must consider the opportunity cost of not developing the land and must be sure that the Nature Conservancy is not restricting development beyond an agreed upon level. These costs of measuring and monitoring private transactions will always be positive, but each party to the contract gains if he or she can reduce them.

Of course, citizens who demand goods and services from government also must measure and monitor the performance of the politicians and bureaucrats supplying them. Like a consumer displeased with food purchased from the supermarket, a citizen who is unhappy with the actions of his political representative has experienced the cost of measuring and monitoring supplier performance. Political outcomes do not always reflect citizens' desires; in the eyes of voters, the political process may, therefore, supply too many of some goods, say nuclear arms, or too few of other goods, say quality education. If sufficiently displeased, he must take action to rectify the problem or at least not support the restaurant or politician in the future.

There are several reasons that these contracting costs are likely to be systematically lower in market processes. While it may seem that self-interested individuals will always cheat if they believe they can avoid detection, there are also incentives for people to resist cheating. For example, people with a reputation for honesty are better trading partners because the costs of enforcing contracts are lower. Reputation capital becomes a valuable asset worth cultivating.

Furthermore, competition among both consumers and suppliers gives each side of the bargain alternate trading partners and therefore discourages cheating on contracts. Contracting costs will not be completely eliminated, but at least competition among buyers and sellers encourages traders to find ways to use brand names or independent rating companies to lower contracting costs.

In the political sector, if a citizen does not believe he is getting from government the goods and services he desires, he can attempt to sway a majority of the voters and elect new suppliers or he can physically move from one location to another. In either case, the costs of changing suppliers are much higher than in the private sector, where there is more competition among potential suppliers. For example, if a local supermarket does not sell what a customer desires, the customer has alternatives from which to make purchases. Even in the more complex case of corporate managers, a stockholder can change agents by selling shares in one company and purchasing shares in another. Simply, because changing suppliers in the private sector does not require agreement from a majority of the other consumers, change is less costly. This condition imposes a strong competitive discipline. In general, information through prices, internalization of costs and benefits from monitoring by individuals, and agent discipline imposed by competition reduce measurement and monitoring costs in market processes.

Measuring and monitoring the actions of political agents can be especially high at the national level.[19] At lower levels of government, the possibility of voting with one's feet and thus changing jurisdictions creates some competition among political regimes and therefore lowers contracting costs in the same way that competition among market suppliers does. Just as competition among firms encourages more attention to consumers and to production costs, competition among political units is more likely to give citizens what they want. But at the national level, the costs of moving to another country are much higher and therefore competition among political entities will be lower.[20] While a free press and free access to governmental information can reduce monitoring costs, the multitude of decisions made at various levels of government and the large number of constituents represented by each political agent continue to keep these costs high. Monitoring costs may also be reduced somewhat if the political or bureaucratic unit supplying a good or service is more dependent on user fees, as it will be more responsive to users. For example, state parks that depend on user fees offer a wide variety of services to visitors at a much lower cost than national parks that depend on general budget appropriations.[21]

Measuring and monitoring a politician's performance is exacerbated by the problem of rational ignorance. That is, on most issues, voters do not bother to become informed because the costs of becoming informed are high relative to the benefits. In the political process, voters ultimately decide who the suppliers will be. In order to make good decisions, however, voters must gather information about alternative candidates or referenda issues and vote on the basis of that information. If an individual takes the time to become informed and votes on what is best for society, he does a service for his fellow citizens. If, however, the voter is not well informed and votes for things that will harm the society, then this cost is spread among all voters. In other words, well-informed voters produce a classic public good, and as with any public good, other voters will be free riders. Many voters will underinvest in becoming informed, thus remaining rationally ignorant. By contrast, individuals in the private sector bear the costs of being informed, but they also directly reap the benefits of good choices and bear the costs of bad ones.

The counterpart to the rational ignorance effect is the special interest effect, in which well-organized special interest groups can lower the cost of information to members and use the size of their memberships to influence how a political representative views an issue. This gives political agents more leeway to respond to the desires of special interest groups whose members care about the specific policies that affect their group and receive low-cost information about policies from their group. Therefore the general public will be rationally ignorant about the specifics of legislation such as the Clean Water Act or Clean Air Act, but environmental groups and regulated companies will be better informed and will lobby for favorable treatment by legislators. If we assume that the political process works perfectly, which is the equivalent of assuming that markets work perfectly, then each opposing side's countervailing powers would internalize the benefits and costs for the decision maker. But in the absence of such perfection, special interest preferences are likely to dominate.

The combination of rational ignorance and the special interest effect explains why legislation can pass that costs each taxpayer a few pennies but provides significant benefits to special interest groups. Because politicians and bureaucrats are rewarded for responding to political pressure groups, there is no guarantee that the values of unorganized interests will be taken into account even if such interests constitute a majority of the population. Most Americans pay marginally higher prices for petroleum products because oil production is prohibited in the Arctic National Wildlife Refuge.[22] Since the cost to each individual is low and the costs of information and action are high relative to the benefits, each person remains rationally ignorant. On the other side, organized groups that favor preserving wildlife habitat in the pristine tundra gain by stopping drilling in the refuge. To the extent that those benefiting from wildlife preservation do not have to pay the opportunity costs of forgone energy production, they will demand more wildlife habitat. In the absence of a perfect political process, we must depend on good intentions to overpower the special interest incentives built into the imperfect system.

This helps explain why we often get perverse environmental results from political action. Government dams have contributed to the demise of salmon and the loss of wild rivers, and logging on national forests has reduced water quality because not all of the costs are borne by the decision makers. The nature of government funding generates another type of third-party effect by concentrating the benefits on special interest groups while diffusing the costs over a large segment of the population. In other words, the political process operates by externalizing costs and internalizing the benefits to special interest groups. When millions or billions of federal dollars are spent cleaning up a Superfund site, the construction companies that do the cleanup will receive tremendous benefits, but the costs will be diffused over a broad taxpayer base. Not surprisingly, campaign contributions from construction political action committees are more highly correlated with the amount spent on Superfund sites than is the toxicity of the site.[23]

Similarly, the uses of public lands are seldom fully paid for by the users, but are covered by general funds collected through taxes. The political agents who supply land for recreation or wilderness must divert it from other uses such as timber production, for which there is an opportunity cost. Neither consumers nor suppliers in the political process, however, directly pay that cost. The bureaucratic manager or politician who does not own the land does not face all the opportunity costs of his decisions. He takes the forgone values into account only if the political process makes him do so. In contrast, private landowners interested in maximizing the value of the resource must take this cost into account in supplying private recreational experiences. Just as external costs result in too much pollution or in overgrazing of the commons, a political process that externalizes costs can result in excessive production of public goods by distorting the real costs of actions to demanders and suppliers.

The potential to concentrate benefits on winners in the political process and to diffuse costs over a broad range of losers obviously has significant implications for how political games are played. As noted above, voting, lobbying, contributing to campaigns, letter writing, and protesting are all examples of actions designed to influence political decisions. In the political process, wealth is often redistributed rather than created, so that, at best, it is a zero-sum game. Unfortunately, as resources are invested in the redistribution, the game becomes negative sum. Economists call this "rent seeking," where "rent" refers to returns in excess of costs.[24] Whether people or groups make large campaign contributions or form voting coalitions, they do so with the expectation of collecting rents that come at the expense of other citizens. In the absence of voluntary exchange, there is no guarantee of net gains from trade in this rent-seeking process; one group's gain is another's loss. Thus the results of rent seeking are in sharp contrast to the potential gains from trade in markets.

FREE MARKET VERSUS POLITICAL ENVIRONMENTALISM

Because traditional thinking about resource and environmental policy pays little attention to the institutions that structure incentives and provide information in the political sector, practitioners often seem puzzled that efficiency implications from scientific management models are ignored in the policy arena. Efficiency is not the direct goal of private-sector decision makers either, but because profits result from decisions that move resources from lower-valued to higher-valued alternatives, there is a tendency toward efficiency in the private sector. The incentive structure in the political sector is less likely to tend toward efficiency because voters are rationally ignorant, because benefits can be concentrated and costs diffused, and because individual voters seldom (probably never) influence the outcome of elections. For these reasons, it is unlikely that elections will link political decisions to efficiency in the same way that private ownership does in the market process.[25]

With private ownership, profits and losses are the measure of how well decision makers are managing. Even where shareholders in a large company have little effect on actual decisions, they can still observe stock prices and annual reports as measures of management's performance. In other words, private ownership gives owners both the information and the incentive to measure performance.

In the political sector, however, similar information and incentives are lacking. Annual budget figures offer information about overall expenditures and outlays, but it is not clear who is responsible and whether larger budgets are good or bad. Even when responsibility can be determined, there is no easy way for a citizen to buy and sell shares in the government. That is why citizens remain rationally ignorant about most aspects of political resource allocation and rationally informed only about issues that directly affect them. The rewards for political resource managers depend not on maximizing net resource values, but on providing politically active constituents with what they want—with little regard for cost. If political resource managers were to follow the efficiency tenets of traditional natural resource economics, it would have to be because there were honest, sincere people (professional managers) pursuing the public interest.

Anthony Fisher has provided perhaps the best summary of how markets and politics should be compared:

> We have already abandoned the assumption of a complete set of competitive markets. . . . But if we now similarly abandon the notion of a perfect planner, it is not clear . . . that the government will do any better. Apart from the question of the planner's motivation to behave in the way assumed in our models, to allocate resources efficiently, there is the question of his ability to do so.[26]

If market transactions fail to encourage good natural-resource stewardship or environmental quality, it is either because the benefits received by the decision maker are low or because the costs incurred are high. For example, suppose a landowner is deciding whether to forgo commercial timber production to enhance an aesthetic quality. If the aesthetic quality involves a beautiful flower garden, a high fence may be sufficient to exclude free riders and capture the full benefits from the product. However, if the tradeoff is between cutting trees and preserving a beautiful mountainside, excluding casual sightseers might be so costly as to preclude capturing a return on production of the view.

Therefore, the key to getting the incentives right through free market environmentalism is to establish property rights that are well defined, enforced, and transferable. Consider each of these elements.

The physical attributes of the resources must be defined in a clear and concise manner if individuals are to reap the benefits of their good actions and are to be held accountable for their bad actions. The rectangular survey system, for example, allows us to define ownership rights over land and clarifies some disputes over ownership. This system may also help us define ownership to the airspace over land, but more questions arise here because of the fluidity of air and the infinite vertical third dimension above ground. If property rights to resources cannot be defined, they obviously cannot be exchanged for other property rights.

Property rights must also be defendable. A rectangular survey may define surface rights to land, but conflicts are inevitable if there is no way to defend the boundaries and prevent other incompatible uses. Barbed wire provided an inexpensive way to defend property rights on the western frontier; locks and chains do the same for parked bicycles. But enforcing one's rights to peace and quiet by "fencing out" sound waves is more difficult, as is keeping other people's hazardous wastes out of a groundwater supply. Whenever the use of property cannot be monitored or enforced, conflicts are inevitable and trades are impossible.

Finally, property rights must be transferable. In contrast to the costs of measuring and monitoring resource uses, which are mainly determined by the physical nature of the property and technology, the ability to exchange is determined largely by the legal environment. Although well-defined and enforced rights allow the owner to enjoy the benefits of using his property, legal restrictions on the sale of that property hinder the potential for trade gains. Suppose that a group of fishers values water for fish habitat more highly than farmers value the same water for irrigation. If the fishers are prohibited from renting or purchasing the water from the farmers, then gains from trade will not be realized and potential wealth will not be created. The farmer will, therefore, have less incentive to leave the water in the stream.

In sum, free market environmentalism requires well-specified rights to take actions with respect to specific resources. If such rights cannot be measured, monitored, and marketed, then there is little possibility for exchange. Garbage disposal through the air, for example, is more of a problem than solid waste disposal in the ground because property rights to the atmosphere are not as easily defined and enforced as are ones involving the Earth's surface. Private ownership of land works quite well for timber production, but measuring, monitoring, and marketing the land for endangered species habitat requires entrepreneurial imagination—especially if the species migrate over large areas.

Free market environmentalism does not assume that these property rights exist or that they are costless to create. Rather, it recognizes the costs of defining and enforcing property rights and emphasizes the role of entrepreneurs in producing new property rights when natural resources and environmental amenities become valuable. Where environmental entrepreneurs can devise ways of marketing environmental values, market incentives can have dramatic results.[27] Entrepreneurs recognize that externalities provide profit opportunities for those who successfully define and enforce property rights where they are lacking. A stream owner who can devise ways of charging fishers can internalize the benefits and gain an incentive to maintain or improve the quality of his fishing stream. The subdivider who puts covenants on deeds that preserve open space, improve views, and generally harmonize development with the environment establishes property rights to these values and captures the value in higher asset values.

The property rights approach to natural resources recognizes that property rights evolve depending on the benefits and costs associated with defining and enforcing rights. This calculus will depend on such variables as the expected value of the resource in question, the technology for measuring and monitoring property rights, and the legal and moral rules that condition the behavior of the interacting parties. At any given time, property rights will reflect the perceived costs and benefits of definition and enforcement. Therefore, the lack of property rights does not necessarily imply a failure of markets, because property rights are continually evolving. As the perceived costs and benefits of defining and enforcing property rights change, property rights will evolve.

This does not mean that there is no role for government in the definition and enforcement process or that property rights will always take all costs and benefits into account. The costs of establishing property rights are positive and can potentially be reduced through governmental institutions, such as courts. Furthermore, because transaction costs are positive, market contracts will not take all costs into account. In the case of water pollution from sources that cannot be identified (with current technology) at low costs, for example, the definition and enforcement of property rights governing water use may be impossible. In addition, excluding nonpayers from enjoying a scenic view may be costly enough that a market cannot evolve under current technologies and institutions. In these cases, there is a utilitarian argument for considering government intervention, but there is no guarantee that the results from political allocation will work any better than a market with positive transaction costs. If markets produce "too little" clean water because dischargers do not have to pay for its use, then political solutions may also produce "too much" clean water because those who enjoy the benefits do not pay the cost.

ADDRESSING THE CRITICS

There are three main critiques of free market environmentalism: free market environmentalism considers only economic values and ignores environmental values; free market environmentalism pays too little attention to the distribution of rights; and free market environmentalism's focus on markets and politics ignores other important allocative institutions.[28]

Which Values: Economic or Environmental?

Because free market environmentalism focuses on human values, it is criticized by those who argue that saving the environment is a moral issue, not an economic one. Philosopher Mark Sagoff puts it this way:

> Lange's Metalmark, a beautiful and endangered butterfly, inhabits sand dunes near Los Angeles for the use of which developers are willing to pay more than $100,000 per acre. Keeping the land from development would not be efficient from a microeconomic point of view, since developers would easily outbid environmentalists. Environmentalists are likely to argue, however, that preserving the butterfly is the right thing morally, legally, and politically—even if it is not economically efficient.[29]

Assuming that property rights to the land in question are well defined and that the environmental values can be captured, Sagoff is correct.[30] Free market environmentalism argues that the willingness of developers to outbid environmentalists tells us which values are higher. This is not to say that moral values have no place in decisions or that moral suasion is not a valuable tool for influencing human behavior. Sagoff further asserts that "environmentalists are concerned about saving magnificent landscapes and species, keeping the air and water clean, and in general getting humanity to tread more lightly on the Earth. They are not concerned . . . about satisfying preferences on a willing-to-pay basis."[31] Turning moral values into political issues and arguing that it is a matter of treading more lightly on the Earth, however, becomes another form of rent seeking, wherein people with one set of moral values get what they want at the expense of others.

Whose Rights?

The second criticism of free market environmentalism is that it pays too little attention to the distribution of rights. The issue here is who has claims over resources and therefore who must pay whom.[32] To the extent that those wanting to save magnificent landscapes and species must pay landowners for those landscapes and habitats, distribution will be important. It is entirely possible that people with environmental preferences will not have enough wealth to act on their preferences. It is here that environmentalists like to take a page from Marx and suggest that "what is important is not the choices people *do* make but the choices people *would* make if they were free of their corrupt bourgeois ideology."[33] By this reasoning, it is easy to say that environmentalists would be willing to pay more if only they had the resources. Of course, this is not verifiable through voluntary trading and thus opens the door for political redistribution.

A related argument is that the distribution of wealth favors people with nonenvironmental preferences over those with environmental preferences.[34] In the case of public lands, making people pay for use of national parks or forests is unfair because it precludes poor people from using the parks. In the case of private land, big corporations already have the rights to use the land, and poor environmentalists cannot afford to purchase these rights from them. In response, there is the empirical question of whether poor people do, in fact, use environmental amenities such as national parks at their current low price. If they do not, what is the justification for subsidizing the environmental amenity for use by the wealthy? Second, because poor people do not have access to many amenities, there may be an argument for redistributing income in their favor, but the redistribution does not have to come in the form of in-kind services from national parks. If they had more income, they could decide how to spend it without subsidizing wealthy park visitors. Finally, is it the case that environmentalists who demand environmental goods and services are poor compared to the rest of the population? A growing body of evidence suggests that the demand for environmental quality is highly sensitive to income and that members of environmental groups have quite high incomes, thus this argument seems tenuous.[35]

Is the Choice Between Only Markets and Politics?

As described above, market processes and political processes are but two alternatives for addressing natural resource use and environmental quality. Even within each of these there are gradations between individual resource owners, corporate owners, town governments, and national governments. It is becoming better recognized that between markets and government are community organizations that can play a role in resource allocation.[36] These might be communities of fishers who regulate access to a fishery[37] or tribal members who restrict access to a grazing common.[38] In either case, how well the institutional arrangement works will depend on its ability to generate information on values and provide incentives for individuals to act on those values. Thought of in this way, free market environmentalism is less about markets and government and more about how various management institutions determine environmental values and how decision makers respond to that information.

CONCLUSION

Which institutional process is more likely to move resources from lower- to higher-valued alternatives is ultimately an empirical question. Traditional natural resource economics has generally concluded that markets do not do very well and that the political process can do better. Free market environmentalism generally comes to the opposite conclusion. As Sagoff argues with regard to markets, such conclusions often turn on the fallacy of disparate comparison:

> A free market with inviolable property rights, low transaction costs, and so on, may, indeed, treat nature better than does an often bumbling and occasionally corrupt bureaucracy beset by special interests. However, this kind of argument . . . commits the fallacy of disparate comparison. It compares what the perfect market would do in theory with what imperfect governmental agencies, at their worst, have done in fact.[39]

Perhaps this is an effective debating tactic, but it is not inherent in the analytical framework described above. Traditional economic analysis stresses the potential for market failure in the natural resource and environmental arena on the grounds that externalities are pervasive. Free market environmentalism explicitly recognizes that this problem arises because it is costly to define, enforce, and trade rights in both the private and political sectors. In fact, the symmetry of the externality argument requires that specific attention be paid to politics as the art of diffusing costs and concentrating benefits. Assuming that turning to the political sector can solve externality problems in the environment ignores the likelihood that government will externalize costs. Just as pollution externalities can generate too much dirty air, political externalities can generate too much water storage, clear-cutting, wilderness, or water quality.

Free market environmentalism emphasizes the importance of market processes in getting more human value from any given stock of resources. Only when rights are well defined, enforced, and transferable will self-interested individuals confront the tradeoffs inherent in a world of scarcity. As entrepreneurs move to fill profit niches, prices will reflect the values we place on resources and the environment. Mistakes will surely be made, but in the process a niche will be created and profit opportunities will attract resource managers with better ideas. Even externalities offer profit niches to the environmental entrepreneur who can better define and enforce property rights to the unowned resource and charge the free rider. In cases in which definition and enforcement costs are insurmountable, political solutions may be called for. Unfortunately, however, those kinds of solutions often become entrenched and stand in the way of innovative market processes that promote fiscal responsibility, efficient resource use, and individual freedom.

Free market environmentalism recognizes that transaction costs are positive under all institutions. The question is which arrangements minimize these costs. Rather than falling into the fallacy of disparate comparison, the challenge for proponents of markets, politics, or other institutional arrangements is to muster the empirical evidence to support their case. Indeed, since the idea of free market environmentalism was first articulated, researchers have been uncovering a growing body of evidence showing the efficacy of market approaches to environmental problems. Let us consider some of that evidence.

CHAPTER 3

FROM FREE GRASS TO FENCES

Roaming across the northern and western borders of Yellowstone National Park into Montana, the bison that spend their summers in the park enter private lands in search of winter grazing.[1] Montana cattle ranchers object to the migration, because many of the bison carry the brucellosis virus, a disease that can infect cows and cause them to abort their calves. Worse yet, if any Montana cattle are infected with brucellosis, the state will lose its brucellosis-free certification from the federal government. Losing this certification drastically reduces the value of all cattle in Montana, because they cannot be easily entered into interstate commerce.

The migration of the Yellowstone bison is like other cases of pollution in which the actions of one party, in this case the National Park Service and other governmental agencies, affect another, in this case Montana cattle ranchers. And like other forms of pollution, this problem results from ill-defined property rights that would hold one party accountable for his or her impacts on another. If Fido digs up your flower bed, your recourse is to find Fido's owner and hold him accountable for the damages. In the brucellosis case, however, the Park Service does not claim ownership of the wild bison and therefore refuses to be accountable for the harm that might be caused to Montana cattle ranchers.

Where property rights, for both the bison and the cattle, are lacking, problems are inevitable because costs are imposed on unwilling recipients. Migrating bison cause harm when they cross into the physical space of others, just as fluid from a hazardous waste site leaves the confines of the dump and pollutes soil or groundwater. With bison and with pollution, making the owner of the pollutant pay for the damage caused would reduce the level of emissions. The risk of brucellosis transmission or hazardous waste pollution under a property rights system of accountability remains unless or until the recipient of the pollutant can enforce his property rights against trespass. Of course, this will never result in zero emissions because it is always costly to define and enforce property rights.

It is interesting to note how the arguments about how to deal with pollution change depending on who is emitting the pollution and who is receiving it. In the case of hazardous wastes and other industrial pollution, environmentalists will argue that all precautions, regardless of cost, should be taken to control discharge, while the polluters will argue that the costs and benefits of control must be considered. In the case of brucellosis, however, the cattle producers call for no discharge, and environmentalists claim that a little pollution is not a problem.

In the context of free market environmentalism, the preferred solution to both migrating bison with brucellosis and hazardous wastes is the definition and enforcement of property rights. Making those who own a hazardous waste site liable for the ooze that may damage the groundwater gives the owners an incentive to take precautions to prevent damages from migrating waste. As we shall see in detail in a later chapter, when Hooker Chemical disposed of its wastes at the now infamous Love Canal in New York, it took great pains to seal the canal to prevent the wastes from leaking into surrounding land and groundwater. But after the local school board purchased the land for one dollar, under threat of condemnation, safety at the disposal site was compromised by political decision makers who allowed the site to be developed against Hooker's warnings. The school board could act less responsibly because it was not clearly liable for its actions.[2]

The same is true in the case of the Yellowstone bison. Because Yellowstone National Park officials are not liable for the potential damages from brucellosis carried by the bison, the responsibility falls on someone else's shoulders. In this case, liability for ownership remains unclear. Montana tried to control the migrating bison once they crossed out of the park by allowing licensed sport hunters to shoot them. Animal rights activists, however, claimed that this infringed on their rights and ultimately pressured the state of Montana to halt its hunting season. Now the state (in cooperation with the National Park Service) tries to haze the bison back into the park, and when unsuccessful, traps them, tests for brucellosis, slaughters those that are infected, and ships those that are not to some Indian reservations. Still the Park Service claims little responsibility for the problem. If it were responsible, it might find more innovative ways of controlling bison movement or eliminating the virus they carry. If the state owned the bison, then it would have the authority to manage the herd size both inside and outside the park and would be responsible for any damage the animals might cause.

Admittedly, it is difficult to imagine a mechanism for establishing clear property rights to the bison, and fencing the perimeter of a 2.2-million-acre national park seems impossible, but this is the essence of the property rights problem. How and when do property rights evolve so that problems such as this are taken care of through bargaining between the parties who are involved? Foresight does not always tell us how this might happen, but hindsight can be instructive. Hence we turn to the frontier of the American West, where the potential for the tragedy of the commons was great, but where property rights evolved sometimes through private efforts and sometimes through state actions. In either case, changing economic and technological conditions generated forces that led to innovative property rights solutions.

THE COWBOY FRONTIER

Imagine yourself in 1840, riding to the top of a divide in what would one day become the state of Montana and gazing on an endless sea of grass. Knowing that eastern markets are hungry for beef, you decide that there is profit potential in grazing cattle on the prairie grass and driving them to eastern railheads. Competition for the grass seems unlikely, since there are similar valleys over almost every divide. But as time passes and as more cattlemen take advantage of the profit opportunities from grazing the prairies, land, grass, and water become more scarce, and an additional cow in your valley reduces the land's grazing potential. Your breeding program has produced a hearty stock that can gain weight on the prairie grass and endure harsh winters, and you do not want other cattle mixing with yours. It is also becoming clear that water will soon become a constraint; if someone else uses the water for irrigation or mining, it will not be available to you.

Common law, as applied in the East, presupposed that cattle would be fenced, making cattle owners responsible for any damages their animals caused. Common law also allocated water to riparian owners in coequal amounts, making no provision for diversion because abundant precipitation made irrigation unnecessary. To western cattlemen, however, establishing property rights according to eastern common law appeared to be as impossible as fencing in the bison at Yellowstone. Stone and timber were not available for fencing, and water was always in short supply and not always available where it was needed. Eastern institutions and ideas simply were not appropriate for the Great Plains. Historian Walter Prescott Webb captured the essence of the problem:

> The Easterner, with his background of forest and farm, could not always understand the man of the cattle kingdom. One went on foot, the other went on horseback; one carried his law in books, the other carried it strapped round his waist. One represented tradition, the other represented innovation. One responded to convention, the other responded to necessity and evolved his own conventions. Yet the man of the timber and the town made the law for the man of the plain; the plainsman, finding this law unsuited to his needs, broke it and was called lawless.[3]

The problems on the American frontier centered on who owned the land, the cattle, and the water. Out of these problems frontier entrepreneurs developed new property institutions to define and enforce rights that improved resource allocation. Those institutions were not perfect, but they demonstrate the potential for innovative solutions to resource ownership problems.

Land

Land on the Great Plains had several characteristics that affected its productive use. The mean average rainfall over much of the region does not exceed 15 inches a year, precluding the use of land for farming as it was traditionally practiced in the East.[4] The forage on the plains was mainly shortgrass, necessitating large quantities of land for each cow. And the lack of trees meant that it was difficult to fence with natural materials. There was little precedent for the type of agriculture that would be used on the Great Plains, so farmers were forced to drastically alter the productive process.

These same characteristics of the land provided the impetus for changing the methods of defining and enforcing property rights. "There was room enough for all," historian Ernest Osgood wrote, "and when a cattleman rode up some likely valley or across some well-grazed divide and found cattle thereon, he looked elsewhere for range."[5] For much of the 1860s and 1870s "squatter sovereignty" was sufficient for settling questions of who owned the land. But the growing demand for land by cattlemen, sheepherders, and farmers eventually increased its value and, therefore, the benefits from engaging in activities to define and enforce property rights.[6]

Initially, settlers attempted to establish some extralegal claims to property, but as Webb described it: ". . . no rancher owned land or grass; he merely owned cattle and the camps. He did possess what was recognized by his neighbors (but not by law) as range rights."[7] Range rights provided some exclusivity over the use of land; but as the population increased, settlement became more dense and land values rose even more. Individuals and groups began devoting more resources toward defining and enforcing private property rights, and early laws provided ways to punish those who drove their stock from the accustomed range. The idea of accustomed right on the basis of priority rights was also reflected in the claim advertisements that appeared in local newspapers. At the time, it was easy for cattlemen to define their range rights: "I, the undersigned, do hereby notify the public that I claim the valley, branching off the Glendive Creek, four miles east of the Allard, and extending to its source on the South side of the Northern Pacific Railroad as a stock range.—Chas. S. Johnson."[8] Such activities could not be enforced in any court of law, but they were inexpensive and they put others on notice that claims existed. As the value of grazing land rose, so did the rate of return on defining and enforcing property rights. To capture these returns, cattlemen organized in groups and used the coercive authority of government to protect their property. By banding together in stockgrowers' associations, cowmen attempted to restrict entry onto the range by controlling access to limited water supplies. These groups also put pressure on state and territorial governments to pass laws that would punish those who drove stock from their "accustomed range." In 1866, the Montana territorial legislature passed a law controlling grazing on public land, and in 1884 a group of cattlemen in St. Louis suggested that the federal government allow the leasing of unclaimed land.[9] Gradually, the West moved toward private property by restricting entry onto land that was once held in common.

The influence of cattlemen and other land associations remained strong until the winter of 1886-1887, "the severest one the new businesses of the northern plains had yet encountered, with snow, ice, wind and below-zero temperatures gripping the area from November to April, in a succession of storms that sent the herds drifting helplessly, unable to find food or water."[10] Thousands of cattle died that winter, and many ranchers went broke and left Montana and

Wyoming to make a living elsewhere. Land values declined, reducing the need to expend resources on enforcing property rights. From 1886 to 1889, membership in the Wyoming Stock Growers Association dropped from 416 to 183.[11] A similar decline was evident in the Montana Stock Growers Association. In his 1887 presidential address, attended by only one-third of the Montana members, Joseph Scott concluded that "had the winter continued twenty days longer, we would not have had much necessity of an Association; we would not have had much left to try to do."[12]

Although the laws and restrictions on land use took ranchers a step toward exclusive ownership, they still did not stop livestock from crossing range boundaries. Only physical barriers could accomplish that, but fences of smooth wire did not hold stock well and hedges were difficult to grow and maintain. The cost in money and time was simply too high. But when barbed wire was introduced in the West in the 1870s, the cost of enclosing land was dramatically reduced. To homesteaders whose land was invaded by cowboys and herds that trampled crops, barbed wire "defined the prairie farmer's private property."[13]

Some stockmen ridiculed the new fencing material, but others saw the advantage of controlling their own pastures. In Texas, for example, "they began buying land with good grass and water and fencing it. In 1882, the Frying Pan Ranch, in the Panhandle, spent $39,000 erecting a four-wire fence around a pasture of 250,000 acres."[14] Other cattlemen enclosed their "accustomed range" with the inexpensive and easy-to-use barbed wire. But a federal law passed in 1885 forbade the fences and provided for the "prosecution of those who stretched fences out upon the public domain."[15] Then, the inevitable conflicts over ownership were settled through both range wars and legal institutions.

Between 1860 and 1900, changing land values and costs of fencing caused individuals and groups to devote more effort to definition and enforcement activity. As a result, the institutions governing landownership on the Great Plains moved toward exclusivity. Measures were enacted that attempted to control grazing on the public domain, and efforts were made to lease unclaimed public lands from the government. During the 1870s and 1880s many acres of land were privately claimed under the homestead, preemption, and desert land laws. Finally, land was granted outright to the transcontinental railroads, which transferred much of it to private hands.[16]

Livestock

As with land, new institutions were also needed in the West for defining and enforcing property rights in livestock. In the East, where farms and herds were much smaller, it was easy for an owner to watch his animals and to know when they strayed from his property. Identifying animals by their natural markings was also feasible on farms that had only a few head of livestock. Furthermore, the lack of common property in the East and the availability of stone and rails for fencing made enforcement of property rights less costly. But a western livestock producer not only had to run his cattle over a large acreage, he also had to pasture them on lands over which he did not always have exclusive control. These factors, combined with the difficulty of fencing large areas where wood was scarce, made eastern methods of enforcing property rights to livestock costly on the plains.

The settlers had to search for alternatives. During the 1860s, sheepmen turned to herding, while "property rights in unbranded cattle were established by the fact that they ran on a certain range. . . ." As long as individuals agreed on who owned the animals, there was little need to devote valuable resources to definition and enforcement questions. Increasing human and cattle populations in the West, however, created more disagreement, and incentives changed. As Osgood described it: "The questions arising over the ownership of cattle and the rights of grazing were intensified as the number and value of the herds increased."[17]

Cattlemen responded by increasing their property rights activities. Although cattlemen had used branding to identify their stock since they had first come to the region, the laws governing branding activity changed.

> There was a time when brands were relatively few and a man could easily remember who owned the different ones, but as they grew more numerous it became necessary to record them in books that the ranchers could carry in their pockets. Among the first laws enacted by territorial legislatures were those requiring the registration of brands, first in counties and later with state livestock boards.[18]

Laws passed in Wyoming and Montana provided for the central registration of distinctive brands, but more laws were needed as the population increased.[19] Osgood captured the effect of this shift on enforcement activity in cattle raising:

> . . . additional laws were passed to further define and enforce rights to cattle. Legislatures passed laws requiring that cattle driven through a territory had to have their brands inspected. Brands were made transferable, and penalties were imposed on those who failed to obtain a bill of sale with a list of brands on the animals purchased. As the complexity and number of brands increased, the resolution of conflicts was turned over from the county clerk to a larger committee including resident stock growers. Laws regulating illegal branding were strengthened by making offenses a felony.[20]

Individual efforts to define and enforce property rights in livestock were complemented by voluntary collective action that gave cattlemen the opportunity to capture gains from economies of scale. Originally, each rancher gathered and branded his own cattle. On the open range, this meant that herds were rounded up as many times as there were individual operators.

As the number of operators increased, however, the cost of handling the cattle in this fashion increased proportionately and cooperation became profitable.

The cattlemen on the plains also used line camps, which were essentially human fences, to enforce their rights to cattle and land. The movie scene of cowboys sitting around a campfire singing songs comes from the camps that were established on the perimeters of range areas. Cowboys spent their days and nights riding the boundaries between ranges, making certain that cattle did not wander from their designated range. The line camps also helped enforce property rights by guarding against rustling. But human fences, while effective, were expensive.

Technology provided the alternative that dramatically changed the face of the American West. In the 1870s, homesteaders and ranchers began using barbed wire to define and enforce their rights to land. By confining cattle to a certain range, cattlemen could reduce both their losses from strays and the costs of rounding up the cattle for branding and shipping. Furthermore, once their cattle were separated from other herds on the range, ranchers could practice controlled management and breeding. In 1874, 10,000 pounds of barbed wire were sold; by 1880, just six years later, over 8.5 million pounds had been sold and the fencing was being used throughout the West.[21]

Most of the changes in defining and enforcing property rights were toward greater exclusivity, but we would expect to find a movement in the opposite direction if asset values declined. An example of this occurred in the 1920s in eastern Montana. From 1918 until 1926, the price of horses in Montana dropped dramatically, from $98 to $29 per head.[22] Prices were depressed as farmers replaced their horses with machines and the U.S. Cavalry stopped buying horses. With the incentive to maintain ownership reduced, many horse owners found it unprofitable to define and enforce property rights to the animals and, therefore, allowed their animals to run the open range. As a result, wild horse herds increased so rapidly that community roundups were held to clear the range of the unclaimed property. Many of the wild horses that cause grazing problems on public lands today are descended from the horses abandoned during the 1920s.

Water

Water presented special ownership problems in the West.[23] Like livestock, water moves freely across many pieces of real estate, but unlike livestock, it is nearly impossible to fence except in reservoirs. Complicating the matter further, the quantity of water can vary from season to season and even from day to day. This is especially true on the Great Plains, where average rainfall ranges from 15 to 20 inches a year. The ever-changing physical nature of the resource makes definition and enforcement of rights to water most difficult. Sir William Blackstone, an eighteenth-century jurist, described it this way: "For water is a moving, wandering thing, and must of necessity continue common by the law of nature; so that I can only have a temporary, transient, usufructuary property therein."[24]

To frontiersmen and settlers on the plains, having access to water was essential. As a result, initial settlement patterns in the region can be traced to river and stream bottoms. During the early years of white settlement, if an individual found a stream location occupied, he simply moved on to another site where there was a supply of water. As long as there was unoccupied land adjacent to streams and as long as the primary use of water was domestic and livestock consumption, westerners found it sufficient to enforce water rights using riparian doctrine from common law, which gave all riparian owners coequal rights to undiminished flows.[25]

As settlement pressure increased, however, and as water was used to irrigate nonriparian land, pressures for changing water institutions grew. Especially in the arid states on the western plains, where water was essential for raising crops or livestock, land with available water became increasingly scarce and the value of water rights rose. In gold mining areas, water was required at the mine site, which was often far from the nearest stream. These conditions induced individuals to devote more resources to redefining property rights in water. In the mining regions, for example, there was no established custom of mining and no recognized law, so miners set up mining districts, formed miners associations, and established mining courts that provided laws.

> These miner's rules and regulations . . . were simple and as far as property rights were concerned related to the acquisitions, working, and retention of their mining claims, and to the appropriation and diversion of water to be used in working them. . . . There was one principle embodied in them all, and on which rests the "Arid Region Doctrine" of the ownership and use of waters, and that was the recognition of discover, followed by prior appropriation, as the inception of the possessor's title, and development by working the claim as the condition of its retention.[26]

Miners recognized the need for an alternative system of water law in the West and worked hard to have California and United States courts recognize their customs and regulations regarding water.[27]

Although precedent established in California in 1850 lowered the cost of establishing property rights in water, the growing scarcity of water on the Great Plains increased the benefits of definition and enforcement activity. Settlers moved toward a system of water laws that (1) granted to the first appropriator an exclusive right to the water and granted to later appropriators rights conditioned on the prior rights of those who had gone before; (2) permitted the diversion of water from streams for use on nonriparian lands; (3) forced water appropriators to forfeit their rights if the water was not used; and (4) allowed for the transfer and exchange of water rights between individuals.[28]

The activities designed to establish and enforce exclusivity were strongest in areas where water was the scarcest. In Montana, Wyoming, Colorado, and New Mexico, where rainfall averages 15 inches per year, the common law of riparian ownership was completely abandoned; where rainfall was greater, in North Dakota, South Dakota, Nebraska, Kansas, Oklahoma, and Texas, states only modified the doctrine.[29] The evolution of water law on the Great Plains was a response to the benefits and costs of defining and enforcing the rights to that valuable resource.

CONCLUSION

Property rights are not static, but evolve through the social arrangements, laws, and customs that govern asset ownership and allocation. As long as the benefits of establishing ownership claims are low relative to the costs, there is little incentive for individuals to define and enforce private property rights. In this case, any tragedy of the commons will be small. As the ratio of perceived benefits and costs changes, however, so will the level of definition and enforcement activity. The higher the value of an asset and the higher the probability of losing the right to use that asset, the greater the incentive will be for institutional entrepreneurs to devise innovative mechanisms for establishing property rights. Technological advances or lower resource prices, which reduce the opportunity costs of definition and enforcement, will increase property rights activity. As open access created conflicts in the American West, individual efforts were channeled toward solving the problems of ownership to land, livestock, and water.

These examples teach us that we should not be too quick to conclude that market-based, property rights solutions will not work for natural resources. Before the invention of barbed wire, fencing vast tracts of land on the Great Plains seemed impossible. For much of the last century, technology for fencing bison, whales, or grizzly bears seemed equally impossible. But as asset values change, so do incentives. Tradeable rights in whale stocks, for example, are highly plausible today, thanks to the global position system (GPS), DNA testing, and radio and acoustical tagging of species.[30]

Similarly property rights in free roaming wildlife, such as wolves, appears at our doorstep. Wolves have been reintroduced into Yellowstone National Park, but ranchers complain that wolves threaten their livelihood by preying on livestock. Could the wolves be fenced? Technology is currently available for fencing dogs by burying a cable that emits a radio signal on the perimeter of a piece of land; the signal, received in the dog's collar, gives the animal a mild shock, which causes it to retreat from the perimeter. Could the same technology be applied to wolves? When red wolves were reintroduced into South Carolina wildlands, they were equipped with radio collars that allow the animals to be tracked. If a wolf wanders too far afield, a radio-activated collar injects the animal with a tranquilizing drug so that it can be returned to its designated habitat. Whales also can be branded by genetic prints and tracked by satellites, providing another way to define property rights.[31]

Lessons from the American West teach us not to underestimate the potential of innovative entrepreneurs to solve problems by establishing property rights on the environmental frontier. If environmental entrepreneurs want to take responsibility for migrating bison or wolves, they should be allowed to do so. For free market environmentalism to be effective, the legal barriers to these innovative solutions must be minimized.

CHAPTER 4

FROM BARBED WIRE TO RED TAPE

with Timothy Iijima

If the nineteenth century was an era of acquisition and privatization of the public domain, the twentieth century was one of massive public reservation. During the first half of the nineteenth century, the federal estate expanded rapidly as states ceded their claims west of the Appalachians and vast tracts were added through purchase or conquest.[1] With the Ordinance of 1785 and the Ordinance of 1787, the original colonies ceded their western lands to the federal government, and the Louisiana Purchase of 1803 enlarged the federal estate by well over 750 million acres—twice the area of Alaska. Because there was no support for leaving the land in the public domain, the government was faced with how to dispose of it. During its early stages, this movement pitted Alexander Hamilton, who favored selling the public lands to enhance the U.S. Treasury and pay off debts incurred during the Revolutionary War, against Thomas Jefferson, who wanted to promote an agrarian ethic by giving the land to those who were willing to settle and cultivate the western frontier. Neither side in the debate questioned the wisdom of privatization. As a result, the first privatization movement, from 1790 to 1920, put more than a billion acres of public land into private ownership.

Although most privatization of federal lands occurred between the Civil War and World War II, new land policies were set in motion that eventually created a federal estate totaling one-third of the nation's land, and public domain totaling more than 40 percent. Part of this dramatic transformation occurred because politically influential conservationists, such as Gifford Pinchot and Theodore Roosevelt, were convinced that private ownership was promoting the rape and ruin of timberlands and, ultimately, a timber famine. Roosevelt warned that "if the present rate of forest destructions is allowed to continue, with nothing to offset it, a timber famine in the future is inevitable."[2] Gifford Pinchot, the first director of the Forest Service, agreed: "We have timber for less than thirty years at the present rate of cutting. . . ."[3]

While the dire predictions never came to pass, the idea that only professional management by public employees could promote good resource stewardship dominated policy debate in the later nineteenth century. The political question shifted from how to dispose of the federal estate to how the federal government could retain and manage it. The answer was to create federal bureaucracies, such as the National Park Service, the Forest Service, and the Bureau of Land Management. Huge quantities of land, timber, minerals, and water became the domain of what Gifford Pinchot called "an elite corps of professionals."

The cutting of large amounts of timber in the Great Lakes region near the end of the nineteenth century is used as evidence that political control of land was necessary to prevent deforestation by private enterprise. Even though timber famines are not of major concern today, conservationists point to this episode as one of the worst environmental disasters of the period and argue that public ownership was the only way to prevent this from continuing. The era of deforestation buttresses the environmentalist's view that short-term profits motivate private interests to destroy valuable natural resources. Is this the lesson from the Great Lakes experience? Was private ownership the problem or the solution?

ALLEGATIONS OF WASTE

Immigrants to North America before the turn of the nineteenth century found an abundance of natural resources. Raw land was free for the taking, and timber was more of a nuisance than an asset. Except for New England, where naval stores (including masts and turpentine) were valuable, native forests were cleared as soon as possible for growing crops. Even before the arrival of Europeans, Indians girdled the trunks of trees to kill them and burned large tracts of forests, and white settlers continued this practice. When the wood was used to build homes, furniture, and tools, the technology for using wood appeared to be quite wasteful to people in Europe who had long found it necessary to conserve wood. The process of hewing a nation from the forest continued until the frontier reached the Great Plains, where settlers confronted fertile but treeless prairies.

Until the settlers reached the Pacific Northwest, the last old-growth forests available to meet timber demands were in the Great Lakes states. During the 1850s, entrepreneurs responded rapidly to price signals from timber markets, and growth of the Great Lakes timber industry typifies nineteenth-century timber exploitation. Within a few decades, small, local mills had grown into an intensely capitalized, highly integrated industry. Within a few more decades, production peaked, and timber production moved to the Northwest, leaving behind a trail of denuded forests and abandoned mills. By 1910, the cycle of rise and decline of the industry was almost complete.[4]

Critics of the nation's logging industry contend that nineteenth-century timber practices were wasteful. With forests in the United States reduced from an estimated 820 million acres at the beginning of the nineteenth century to 495 million acres in 1932, critics like Gifford Pinchot—who was trained in Germany, where timber endowments were very different—saw the extensive use of virgin forests as wasteful and unscientific.[5] These critics referred to the logging practices around the Great Lakes as the "Great Lakes tragedy." Andrew Rodgers, a historian of American forestry and plant sciences, wrote: "It was assumed that the continent's forest resources were inexhaustible."[6] In his economic history of Wisconsin, Robert Fries warned: "One must not let hindsight obscure the fact that the very immensity of the forests led most people to take them for granted, much as they did the sunshine and the air about them."[7] Bernhard Fernow, the first chief of the United States Division of Forestry, wrote in 1902:

> The natural resources of the Earth have in all ages and in all countries, for a time at least, been squandered by man with a wanton disregard of the future, and are still being squandered wherever absolute necessity has not yet forced a more careful utilization.
>
> This is natural, as long as the exploitation of these resources is left unrestricted in private hands; for private enterprise, private interest, knows only the immediate future—has only one aim in the use of these resources, namely, to obtain from them the greatest possible personal and present gain.[8]

These charges suggest that greed clouded timber managers' perception of resource endowments and led to an unwise or inefficient rate of exploitation.

Some critics charged that additional waste came from logging operations and timber processing. Frederick Merk, for example, observed that "it has been estimated that not more than 40 percent of the magnificent forest . . . ever reached the sawmill." Fries claimed that "a billion more board feet could have been produced in the years 1872-1905 had band saws been used to the exclusion of muley and circular saws." Conservationists who desired public reservations also alleged that the timber industry thrived because it was allowed to trespass on public timberlands: "By 1850 trespassing had become an accepted practice in the Great Lakes pinery. . . . To take trees from the public domain was no more immoral than it was to float a canoe on a public river." One source estimated that between 1844 and 1854 nearly 90 percent of the 500 million board feet shipped from eastern Wisconsin and northern Michigan had been stolen from the public domain.[9]

EVIDENCE TO THE CONTRARY

The charges of waste are taken at face value by environmentalists who criticize private resource owners, but evidence takes on a different meaning when examined under the lens of free market environmentalism. In particular, we must ask what constitutes waste given the incentives and information facing people at the time. There was no question that nineteenth-century loggers used techniques that would be considered wasteful today, if wasteful means loggers threw away wood that would be processed and used today. But if wasteful means that loggers used economically inefficient practices or that the value of the wood that was thrown away was greater than the cost of conserving it, then the extent of waste is much less clear.

Environmentalists' objections to private ownership are buttressed by examples like one from the 1880s in which a logging firm allegedly stole approximately 1.25 million board feet of timber from someone else's land, over a million board feet of which was cut simply to get it out of the way so the best trees could be harvested.[10] Of course, if the timber was stolen, there was no guarantee that efficient forestry practices took place. Therefore, this incident should not draw criticism of private ownership, but criticism of the way property rights were enforced, which is the domain of government.

Furthermore, even if the timber was not stolen, we cannot be certain that its management was wasteful in the context of the relative scarcities of nineteenth-century labor. Labor was scarce and expensive and resources were abundant and inexpensive.[11] This was particularly true in the lumber industry. Merk wrote: "The remoteness of the lumber camps from settlements, the rough and temporary nature of the work, and the unsatisfactory terms of employment were sufficient . . . to render the labor problem in the pineries a troublesome one."[12] Given the relative prices of labor and time, conserving abundant trees by expending expensive labor would have been wasteful. "It is to be noted as a characteristic of all sawmill innovations of this day that they were calculated solely to secure an increased output or a saving of labor. Little effort was made toward effecting a saving of lumber since timber was still cheap and abundant."[13] In short, waste and efficiency are not absolute concepts measured by energy input and output; they must be considered in the context of the relative economic scarcity at the time when decisions were made.

The allegation that firms operated under a delusion that timber stands were inexhaustible requires a consideration of the information and the incentives timberland owners and managers had about present and future supplies and demands. Those firms that understood how increasing scarcity resulted from decreasing supplies and increasing demand stood to profit from the knowledge. Moreover, the allegation that firms failed to adequately conserve timber for the future fails to consider the economics of resource allocation over time.[14]

In the long run, a firm can remain profitable only if it accounts for both present and future demand. Firms that recklessly exploit a resource for the quickest possible financial gain, whether because of ignorance or carelessness, only make it more profitable for informed firms to purchase and hold large amounts of the resource off the market until its value rises as a result of current exploitation. An informed owner of resources who notices that much of the supply is being squandered and sold for immediate gain will anticipate an increase in prices. In this way, the nation's future demands for timber would be efficiently accounted for if only a few firms foresaw Pinchot's "timber famine."

In the case of timberlands, prudent investors would forgo cutting only if they expected the value of standing timber to increase faster than the return on alternative investments. Typically, forestland was inspected, appraised, purchased, and sold to lumber mills, activities that depended on current and future prices. The decision for an owner was whether and for how long to hold timber and at what rate to harvest it. If a private owner with secure property rights expected to earn a higher rate of return from harvesting the asset and investing the proceeds elsewhere, then holding onto timbers would be a losing proposition. In this case, the value-maximizing decision would be to sell the timber (presumably to be cut and processed) and put the proceeds into stocks, bonds, or other investments. Alternately, a timber owner would be foolish to sell trees that are expected to increase in value at a rate higher than the prevailing return on other investments. Through this process, the long-run prevailing interest rate serves as a guide for determining the rate at which timber resources would be harvested. In other words, if timber owners believed the predictions of a timber famine, they surely would have expected future timber prices to be significantly higher and would have profited by saving the trees for later harvest. Their behavior, however, suggests that they did not believe Pinchot's predictions.

Economic reasoning implies that under efficient management, timber prices should have been steadily increasing at the prevailing interest rate.[15] If timber prices were not rising at the prevailing interest rate, it would have been more efficient and profitable to liquidate the trees at a faster pace, investing the proceeds in other assets. Conversely, if timber prices were rising faster than the return on other assets, it would have made sense for private owners to reduce their harvest rate, thereby retaining an investment in growing timber. The early conservationists' allegation that timber was being harvested too fast suggests that current prices were being suppressed and that once the timber famine was realized, there would be a price shock similar to the one that occurred during the energy crisis in the 1970s.

Any price shock caused by an expected timber famine would have been exacerbated if the timber companies had been stealing trees from government lands rather than legitimately cutting from their own property. In this case the harvest calculus is very different because unowned trees are subject to the tragedy of the commons. In the commons, anyone leaving a tree for future harvest will only find that someone else takes the tree first.[16] With the tragedy of the commons, it is reasonable to expect rapid harvesting to keep the price of lumber low until the free trees are depleted, at which time the price will jump to reflect the price that lumber mills will be forced to pay for scarce private timber.

Certainly, some timber theft did occur before 1860, but only because by that time the government had just begun making the public domain available for private ownership. When the majority of good forestlands were publicly owned and few forestlands were open to private ownership, the industry had no choice but to take public resources.[17] It is to the government's credit, however, that little time was wasted in selling off land in the Great Lakes region; by privatizing land, the government allowed the market to allocate the resources according to scarcity rather than on the first-come/first-served basis of theft.[18]

Evidence that Midwest timber prices steadily increased and were free of price shocks suggests that timber theft did not interfere significantly with efficient timber allocation and that timber markets were taking account of long-run timber supply and demand. Economists Ronald Johnson and Gary Libecap examined the time path of timber prices and found no discontinuity. To the contrary, they found a steady and smooth rise in nineteenth-century timber prices. Johnson and Libecap examined stumpage prices (the price of standing timber) and reported that rates of return were relatively constant through 1900, with the rate of change in stumpage prices hovering at around 6 percent. This is within approximately one percentage point of the prevailing yield on railroad bonds over the same period of time.[19] Their evidence suggests that timber owners were acting rationally in taking a long-run view of resource allocation.

Data on the price of processed wood yield a similar conclusion. Warren and Person report "a steady increase in the purchasing power in the price of lumber from 1789 to the present time. From 1798 to 1914, the purchasing power of lumber increased at the compound rate of 1.54 percent per year."[20] The rise in price was steady and absent of price shocks, suggesting that neither the legend of inexhaustibility nor timber theft was significant enough to affect the market price of timber.

There are several important consequences that result from this efficient market. As prices gradually rise, firms gain an economic incentive to find new and less expensive alternative resources. Firms that find less expensive alternatives to higher-priced timber will profit as the consuming public switches to the alternatives; firms that do not seek those alternatives risk losing a significant share of business to those that do. This explains why firms began developing production in the timber stands on the West Coast and in the South, where tremendous amounts of standing timber were available.[21] Furthermore, because the price of timber will rise gradually, transition to a different source is likely to be phased in gradually. In these ways, a stable supply of resources is assured for the economy.

As these effects on the supply are taking place, the economy's demand for the diminishing resource is scaled back as consumers find it financially important to conserve, to develop more efficient ways of using the resource, and to substitute other materials. During the late nineteenth century, for example, the railroads foresaw increasing scarcity and invested in research on protecting wooden bridges and ties against decay, on the potential for using steel to make ties and bridges, and on the science of efficient bridge architecture.[22] The shifts from wood to coal to diesel as sources of fuel provide another example of price signals at work.

Land speculation provides additional evidence that private landowners were cognizant of the long term. Though many historical accounts of nineteenth-century land policy lament the way speculation held land and timber off the market and delayed development, such speculation clearly indicates the power of market forces to conserve resources.[23]

Beginning in the 1850s, speculation on timberland in the Great Lakes region became big business. When tracts were opened for bidding, speculators assessed the timber potential in order to make informed bids. They also gathered information on the general conditions of resource scarcity so they could estimate the future value of the trees they were purchasing. During the years when federal lands were being sold in the Great Lakes region, annual land claims rose in proportion to annual lumber production, and private ownership was secured well before major harvesting began in an area.[24]

Philetus Sawyer, a lumber tycoon and United States senator, attributed his fortune to early, well-placed purchases of forestlands, some of which he kept off the market for as much as a quarter century. Sawyer profited because he went to the expense of investigating the quality of forestland, about which little was known, and because he invested capital in trees whose value would not be realized immediately.[25]

Historian Paul W. Gates has described one of the most successful land speculation projects in American history. As a land grant college, Cornell University obtained rights to nearly a million acres from the federal government in the 1860s. It claimed approximately one-half million acres in Wisconsin, managed the lands, and gradually sold them. Most of the land was sold by 1890, although final sales were not completed until 1925. Cornell hired a talented and well-placed land agent who meticulously chose stands and skillfully negotiated the timber sales to lumber mills. By carefully controlling the timing of timber and land development, Cornell sold lands that were initially worth $5 per acre for $20 per acre; the richest tract brought $82 an acre. Gross revenues minus gross expenses from Cornell's land sales amounted to approximately $5 million.[26]

This does not mean that private entrepreneurs did not cut millions of trees and leave some land denuded. The Kingston Plains offers a pointed example. Unlike most land in the Great Lakes area, which is now productive farmland, timberland, or recreational land for fishers, hunters, and hikers, the Kingston Plains has never recovered from logging done a hundred years ago. Efforts have been made to replant the area, but the soil is too infertile and sandy. It took hundreds of years for original forest to grow, and it will take hundreds of years for the area to recover.[27]

Again, however, the decision to clear-cut the area probably made very good economic sense. When the trees were cut, good timber stands in the Great Lakes area were selling for around $20 per acre.[28] In order to determine whether it would have made more sense to invest in trees by forgoing the harvest, we must consider the return on other investments. Had the income from selling these trees been invested in bonds or some other form of savings at the time, it would now be worth approximately $110,000 per acre, or $2.8 billion for the 40 square miles.[29] If the trees at Kingston Plains had been left standing, would the benefits derived over the past one hundred years from preserving land for wildlife habitat, hiking, and other environmental amenities have been worth forgoing the benefits society received from logging? The answer is highly subjective, but the tremendous benefits from exploiting the Kingston Plains cannot be ignored. Because the land in this area is not worth anything close to this, we must infer that harvesting the trees was the correct economic choice.

COMMODITY EFFICIENCY OR AESTHETICS?

Most of the early conservationists' arguments were concerned with timber production and timber famine and not wilderness or aesthetic values common in today's policy debates. No one would deny that most of the old-growth timber was harvested in the Great Lakes region, but wilderness was not a scarce commodity at the time. Furthermore, except for a few isolated areas where poor soil quality has retarded reforestation, most of the denuded lands have recovered to become productive timber or agricultural lands.

This leaves the question of whether the Great Lakes timberlands were wasted in terms of aesthetic or environmental values that are not included in commodity considerations. We must keep in mind that decisions, whether in the private or the political sector, are always made in the context of contemporary values and information. In 1900, per capita income was one-tenth of what it is today, and most of the population was not wealthy enough to demand aesthetic values. Backpacking, canoeing, fishing, and hunting were not leisure activities; they were means of transportation or food production. The untamed wilderness—where today we look for peace and quiet—was a nuisance and a source of danger. One writer observed in 1857:

> The lumbermen on the Upper Wisconsin, are not only men of means to prosecute their business with eminent success, but they have the further qualifications of intelligence, energy, and perseverance. . . . the proof is in the reduction by them, in a few short years, of those wild wastes, into a land of productive industry, equaled by no other in the state—scarcely in the west.[30]

Although the conservation movement was gaining strength as logging activity progressed through the old-growth forest of the Great Lakes region, there is solid evidence that these logging activities were widely viewed as beneficial. For example, "the idea of conservation seems not to have taken hold of the people of Minnesota during the years when these natural resources were being harvested so rapidly."[31] Frederick Merk concluded that "the swift forest destruction that accompanied the expansion of the lumber industry gave concern only to a few obscure idealists."[32]

Conservation legislation moved forward only in small steps and against strong opposition. In 1910, Edward Griffith, Wisconsin's first state forester, urged that counties be empowered to acquire forest in cooperation with the State Board of Forestry. The public responded with open hostility. The county board of Oneida County referred to it as "Mr. Griffith's pet scheme to gobble up our best agricultural lands." Malcolm Rosholt wrote: "In short, it was still the will of the people, even though most of the pine timber was already gone, to do nothing towards reforestation or towards calling a halt to indiscriminate timber cutting, while encouraging cultivation of land no matter how sandy, stony, swampy, or unsuited to farming it was."[33]

The economics of nineteenth-century timber harvesting, however, did not always work against the environment. In fact, the remaining virgin stands in the Great Lakes area have been saved largely by market forces. Where timber stands are largely inaccessible, difficult to log, or sparse, the timber was left generally untouched throughout the nineteenth century because logging those areas would not have been profitable. In northern Michigan and parts of Canada, thousands of acres were spared because the cost of logging them was simply not worth the value of the timber. Examples of such land include several thousand acres in northern Michigan, smaller portions of the Boundary Waters Canoe Area in northern Minnesota, and numerous other scattered old-growth stands in the northern Midwest.[34] Thus, economic realities aided in preserving many trees.

As incomes rose and people became more interested in leisure, recreation, and the environment, private individuals started purchasing and preserving timber stands. Two examples of private land conservation that were donated to the U.S. Forest Service are the 17,000-acre McCormick Tract, originally owned by Fisher Body Company, and the 21,000-acre Sylvania Tract, once the private retreat of Cyrus McCormick. Another example of old-growth forest still privately managed is the Huron Mountain Club.[35] Located on Michigan's Upper Peninsula, the Huron Mountain Club is the brainchild of Horatio Seymour, Jr., a manager of the Michigan Land & Iron Company. It was established on November 29, 1897, with 50 members, each of whom was expected to contribute $5,000 to a capital fund, and a 12-member board to oversee club business. The club initially acquired 7,000 acres by purchase or lease, and today has about 25,000 acres under its control.[36] In the early years when the board of directors periodically decided to cut timber to raise capital for improvements or land acquisition, it did so selectively in the exterior region of the property. They always left 6,000 acres in the interior untouched as a reserve. In 1937 the club enlisted the services of Aldo Leopold, the nationally renowned conservationist, to assess the natural and scientific values of their land and to recommend ways to protect those values. After surveying an array of flora and fauna, Leopold concluded that "the Huron Mountain property would soon be one of the few large remnants of maple-hemlock forest remaining in a substantially undisturbed condition."[37] Since Leopold's assessment, the Huron Mountain Club has continued to acquire land and to actively manage it for recreation and biological diversity.[38]

Farther west, private commercial interests were also interested in conserving natural amenities. Contrary to the popular myth that national parks were established by the federal government at the suggestion of a few farsighted, unselfish, and idealistic men who foresaw the national need, Yellowstone National Park and its major counterparts were the product of transcontinental railroads that saw the profit opportunity in delivering passengers to the areas. In the case of Yellowstone, our first park established in 1872, the Northern Pacific Railroad was the driving force, funding early expeditions and lobbying Congress for setting aside the reserve. As one railroad official put it,

> We do not want to see the Falls of the Yellowstone driving the looms of a cotton factory, or the great geysers boiling pork for some gigantic packing house, but in all the native majesty and grandeur in which they appear today, without, as yet, a single trace of that adornment which is desecration, that improvement which is equivalent to ruin, or that utilization which means utter destruction.[39]

To prevent the area from being homesteaded, mined, or logged, and hence the amenities destroyed, the officials of the Northern Pacific lobbied for national park status. It commissioned photographer William Henry Jackson and painter Thomas Moran to document Yellowstone's beauty and natural wonders and used their work in the lobbying process by placing Jackson's photographs on the desk of every member of Congress and some of Moran's watercolors on the desks of the most influential senators and representatives. After the enabling legislation passed quickly with only token opposition, it paid for wardens to patrol against poaching and built roads and hotels to serve visitors. By establishing a monopoly on internal services, the Northern Pacific was able to capture most of Yellowstone's amenity value through train fares to the park, transportation within the park, and meals and lodging at its hotels. To be sure, railroad officials were motivated by profits, but their actions resulted in the preservation of cornerstones of the national park system, including Glacier (the Great Northern Railroad), Mount Rainier (the Tacoma Eastern Railroad), Crater Lake (the Southern Pacific Railroad), and the Grand Canyon (the Santa Fe Railway).[40]

CONCLUSION

The evidence on Great Lakes timber production during the late nineteenth century presents no justification for the massive reservation of timberlands implemented by President Roosevelt and Gifford Pinchot at the turn of the century. When judged against prudent investment criteria, nineteenth-century timber markets were taking account of long-run timber supply and demand. Even if nonmarket, noncommodity values are included in the calculations, it is not clear that markets incorrectly accounted for these values. And considering the contemporary values of nineteenth-century citizens, it is difficult to argue that even the most omniscient decision maker would have behaved differently. It is not clear that transferring management of the Great Lakes timberlands to the political arena would have or should have done anything to alter resource allocation.

Nevertheless, Roosevelt and Pinchot won the battle in the political arena. From policies of disposal of the federal estate, we moved to policies of reservation and political management. The establishment of Yellowstone National Park in 1872 and the extensive national forest reserves in 1891 marked a significant change in the role of the federal government in resource management. With over one-third of the nation's land (in the lower 48 states) now under control of the federal government and another 7 percent under control of the states, it is vital that we critically examine the results of political control.

CHAPTER 5

BUREAUCRACY VERSUS ENVIRONMENT—THE BEAT GOES ON

Although efforts to reserve millions of acres in the political domain were under way during the late nineteenth century, Mr. and Mrs. W. W. Beck were doing their part to preserve one little corner of the world on the outskirts of Seattle, Washington. In 1887, they bought several parcels of land with giant fir trees reaching 400 feet in height and 20 feet in diameter. The Becks built a pavilion for concerts and nature lectures and added paths, benches, and totem poles. Ravenna Park soon became immensely popular. Visitors paid 25 cents a day or $5 a year ($3.60 and $72, respectively, in 1999 dollars) to enter the park. Even with the fees, 8,000 to 10,000 people visited the park on a busy day.[1]

As the Seattle population grew and conservationist sentiment developed, residents began to lobby for acquiring more public parklands, including Ravenna Park. In 1911, the city bought Ravenna from the Becks for $135,663 following condemnation proceedings. Shortly after the city's acquisition, according to newspaper accounts, the giant firs began disappearing. The Seattle Federation of Women's Clubs confronted Park Superintendent J. W. Thompson with reports of tree cutting. He acknowledged that the large "Roosevelt Tree" had been cut down because it had posed a "threat to public safety." It had been cut into cordwood and sold, Thompson conceded, but only to facilitate its removal and to defray costs. The federation asked a University of Washington forestry professor to investigate. When the women brought the professor's finding that a number of trees had been cut to the attention of the park board, the board expressed regret and promised that the cutting would stop. By 1925, however, all the giant fir trees in Ravenna had disappeared.[2]

Some people still blame the destruction of the trees on a 1925 windstorm; others blame it on automobiles and chimney smoke. But it was the bureaucracy that destroyed what the Becks had saved. Park employees took advantage of their access to the park and cut down trees to sell firewood. Park Department records charge Superintendent Thompson with abuse of public funds, equipment, and personnel, plus the unauthorized sale of public property. Even if he and his subordinates were not direct culprits, they had allowed the cutting to go on.

The Ravenna Park debacle occurred at a time when leaders of the early conservation movement were touting public ownership as the only way to conserve America's natural resources. In their view, the greed of private owners was an insurmountable obstacle to conservation. Yet, in the case of Ravenna Park, private owners protected natural treasures, while public agents destroyed them. Even an outcry from public watchdogs could not prevent the eventual destruction of the giant fir trees.

Could similar controversies and results occur with political resources today? Unfortunately, the answer is yes, and on a much larger scale than the incident at Ravenna Park.

TIMBER BEASTS VERSUS TREE HUGGERS

The U.S. Forest Service, with an annual budget approaching $3 billion and over 38,000 full-time employees, is the largest natural resource agency in the federal government. It oversees more than 191 million acres of national forests and is required by law to manage its lands for multiple uses, which include timber production, livestock grazing, mineral and energy production, fish and wildlife habitat, wilderness protection, and public recreation. While the agency does collect revenue from the sale of timber, grazing allotments, minerals, oil and gas, and certain recreational activities such as camping, it is not required to generate a profit. In fact, most of the revenues it collects go to the U.S. Treasury, and expenditures by the agency are appropriated by Congress. Not surprisingly, the agency regularly runs huge deficits. Over the 1994-1996 period, for example, the Forest Service spent an average of $3 billion per year, while receiving $914 million in revenue from timber, recreation, grazing, minerals, and other uses.[3] Recreation incurred the largest average annual loss of $355 million, followed by timber with a loss of $290 million, and grazing with a loss of $66 million.[4]

Though recreation generates the largest deficit, timber production remains the most controversial activity on national forests. Over the past three decades, environmentalists have argued that the Forest Service overemphasizes this activity at the expense of environmental amenities. Moreover, they contend that because timber production loses millions of dollars each year, it should be eliminated. Timber interests counter that the agency gives wilderness values too much attention and that the Forest Service loses money selling timber because the agency has been inundated with costly regulations and administrative red tape that have driven up program costs.

The fact that the Forest Service loses millions selling timber indicates that something is drastically wrong with the program. (The same can be said for the other programs that lose money, such as recreation.) But it is insufficient to conclude that the program should be eliminated. The appropriate fiscal question is: Does the program have the potential to make money for the taxpayer?

Consider national forests in Montana, where federal timber sale programs frequently lose money. Comparisons with state-run forests in Montana indicate that selling timber from national forests in the state does not have to be a money-losing proposition.[5] Montana state and national forestlands are often adjacent to one another, and surveys by silviculturists have shown that these lands have essentially the same timber-growing potential. Yet Montana's state-owned forests netted more than $13 million selling timber over a five-year period (1988-1992). Over the same period, the ten national forests in the state lost nearly $42 million selling timber! Data on state timber operations in Washington, Oregon, and Idaho show that Montana is not unique.[6]

These dramatic differences do not originate on the revenue side of the equation, but are due to differences in management costs. Not only are state and national forests located in the same land base, both are multiple-use forests. And state foresters must carry out the same duties as Forest Service personnel, including overseeing road building and reforestation and conducting environmental analyses. Yet state foresters operate at much lower costs than the Forest Service—and not because they are paid less. The Forest Service uses over twice the number of personnel as do state foresters in preparing and administering timber sales for a given volume of timber harvested. State forests also have much lower road building costs because they build relatively inexpensive roads that last only as long as the logging proceeds, in contrast to the national forest's permanent road system. Hence, the state's roads average about $5,000 per mile in construction costs, while the Forest Service spends at least $50,000 per mile.[7]

What explains the fiscal difference between state and national forest timber operations? In a word, incentives. By law, state foresters must maximize returns from their lands for public schools. Their jobs are on the line if they do not do so. Forest Service personnel have no such requirement. If they lose money selling timber, taxpayers make up the losses. In other words, Forest Service personnel have little incentive to control costs or maximize the net value of the timber they sell.

Further indication that the Forest Service operates under perverse economic incentives comes from a recent review of timber harvests and agency spending. From 1991 to 1996, timber harvests, still the dominant source of revenue, fell from an annual average of over 8 billion board feet to less than 4 billion board feet.[8] Yet, Forest Service annual spending changed very little over this period, hovering around $3 billion per year. More puzzling is the fact that while timber production declined 69 percent from 1981-84 to 1994-96, the share of the Forest Service budget devoted to timber actually increased 10 percent. Meanwhile, the proportion of spending devoted to recreation increased only 3 percent, while recreational use on national forests increased over 50 percent. Recently, the fastest growing segment of the Forest Service budget was general administration, increasing from 19 percent of total expenditures in 1992 to 32 percent in 1996.[9]

Perhaps the pain of the Forest Service's poor economic performance would be reduced if national forests were in a state of ecological health, but perverse economic incentives also lead to perverse ecological incentives. An environmental audit, ordered by the Montana state legislature and conducted by professionals (including environmental-group representatives) reviewed forest practices in Montana in 1992 to ascertain how well foresters protected watershed from the impacts of logging. The audit found that the state ranked higher than the Forest Service.[10] Similarly, an inquiry by the General Accounting

Office into U.S. Forest Service and Bureau of Land Management planning for the period from February 1988 to August 1990 found that the two federal land agencies were not meeting objectives for sustaining wildlife populations.[11] The inquiry found that only one-third of the 51 wildlife plans containing 1,130 wildlife-related actions had been completed.

Equally disturbing is the increasing risk of catastrophic fires occurring on many national forests. Decades of fire suppression has led to a huge buildup of fuel from understory growth and changed the species composition of many national forests. According to U.S. Forest Service fire expert Steve Arno, once-open ponderosa pine forests in the Intermountain West have given way to densely growing stands of Douglas fir because, unlike ponderosa pine, fir is shade tolerant and is able to grow in the thick forest understory that now characterizes many national forests in the region.[12] These stands of fir are also very susceptible to insect infestation. The result is a huge buildup of diseased and dying trees, adding further to the fuel load.

And the fire condition in the Intermountain region is only the tip of the iceberg. Forest Service chief Michael Dombeck recently testified before Congress that nearly 40 million acres, or 21 percent of all national forestlands, are classified as at high risk of catastrophic fire.[13] With the recent steep decline in timber harvests in the 1990s, the risk of fire on national forests can only increase.

The fiscal and environmental problems that plague our national forests are the results of institutions, not people. To expect forest managers to set aside self-interest and objectively weigh the benefits and costs of multiple-use management is to ignore the information and the incentives that confront them. Forest managers are not supposed to manage the national forests to maximize returns, but they can and do manage in ways that maximize budgets at the expense of taxpayers without ensuring environmental quality. Bureaucrats in general have a propensity to expand their staff and budgets because such expansion provides higher salaries, more prestige, and more power. In the case of the Forest Service, timber revenues are generated when trees are harvested, with a percentage of the receipts retained by the bureaucracy. In addition, timber harvests mean a larger road-building staff and more budget for timber management. With little revenue generated from recreational or environmental amenities and less staff required to manage wilderness, these values have received less attention than traditional commodities.

Organized wilderness advocates, however, are increasing pressure for making larger expenditures on recreational and environmental amenities and devoting more land to these uses. Even if the pendulum swings in this direction, controversies will not disappear. With the costs diffused among all taxpayers and with benefits concentrated, environmentalists consider no price to be too great for saving wilderness. For example, when a group of environmentalists and local landowners discovered that a timber company was planning to cut trees from four sections (2,560 acres) in the Greater Yellowstone area, they asked Congress to buy the property for $800 per acre, an amount the landowner was eager to accept, since timberlands in the area are worth only approximately $500 per acre. Such pressures will only increase Forest Service deficits and fuel a backlash from commodity interests whose jobs depend on the forests. The bureaucratic process does little to encourage either fiscal or environmental responsibility.

PARK UPKEEP IN PERIL

Fiscal and environmental controversies surround the National Park Service as well. The mission of the Park Service is "to promote and regulate the use of the . . . national parks . . . which purpose is to conserve the scenery and the natural and historic objects and the wild life therein and to provide for the enjoyment of the same in such manner and by such means as will leave them unimpaired for the enjoyment of future generations."[14] But national treasures such as Yellowstone National Park are far from "unimpaired for the enjoyment of future generations." Sewage treatment systems in the park are so old that they cannot handle the wastes generated during peak visitation periods.[15] Recent monitoring reveals that treated wastewater is contaminating waters that feed Old Faithful as well as popular trout streams, such as Iron Springs Creek. At times raw sewage is actually dumped into meadows as an alternative to dumping it directly into streams and lakes. The park's road system is also crumbling. In August 1998, a part of Grand Loop Road, a popular road in Yellowstone that provides opportunities to view wildlife such as grizzly bears, had to be closed because it posed a hazard to vehicles and threatened passenger safety.[16] The road, which was last surfaced in 1942, is presently inundated with potholes. Though the Park Service spent some funds in 1998 from higher entrance fees to patch and fill the potholes, these repairs failed to last through the summer.

In his book, *The National Parks Compromised,* former director of the National Park Service James Ridenour recognized the impact of "park barrel politics" on the nation's treasures. He states that "the government has just not taken care of these beautiful treasures" and goes on to describe a visit to Sequoia National Park when he was director. "I noticed water running down the pavement and upon closer look, I noticed toilet paper. The old sewer system was overloaded and the pipes were clogged up. I couldn't believe it—here we were trying to set the environmental standard for the nation and we were in blatant violation of the standards ourselves."[17] Writing in *National Geographic,* John G. Mitchell said, ". . . parks were created 'for the enjoyment of future generations.' For many, that 78-year-old promise is eroding."[18]

The Park Service wants more money from Congress for park upkeep, but reports reveal that the Park Service may not be wisely spending the funds it has. For example, the agency spent $330,000 in 1997 building an outhouse complete with two composting toilets and state-of-the-art foundation to withstand earthquakes in a remote area of the Delaware Water Gap National Recreation Area in Pennsylvania.[19] The cost seems exorbitant, given the fact that there is no running water and the facility has to be shut down during winter because the two composting toilets will not work in freezing temperatures. The agency is also building two other, more expensive outhouses in Glacier National Park.[20] Designed by six architects and engineers employed by the Park Service, the multimillion-dollar structures will serve only a few thousand of the two million people who visit Glacier each year. Typical of government projects, this project is loaded with high overhead. By the end of 1997, the project incurred $860,000 for the cost of "design and construction supervision teams" in regional offices in Denver.

Not all of the blame can be laid at the doorstep of the Park Service because politicians with high seniority direct money toward pet projects instead of needed park infrastructure at Yellowstone. One high-ranking Pennsylvania lawmaker garnered $8.3 million in land acquisition and construction money for the Delaware Water Gap National Recreation Area in his district. The park, site of the infamous outhouse, received more money for new buildings and trails than Yellowstone, Grand Teton, and Glacier combined. Recently retired Park Service director Roger Kennedy said more money is spent on "congressionally identified" projects than on agency-recommended projects.[21] Crown jewels such as Yellowstone, however, often lose in the battle for congressional pork because congressional delegations from low population states have less clout than their high population counterparts.. Journalist Frank Greve notes, "The Old Faithful area . . . suffers from a leaky, overwhelmed World War II–era sewage treatment plant" because with two low-seniority Wyoming delegation lawmakers, Yellowstone's construction problems are not high priority on Capitol Hill.[22]

An impaired Yellowstone National Park stems from the fact that about 90 percent of the Park Service budget comes from taxes by way of congressional appropriations. Under this financing approach, politicians who hold the purse strings are able to direct major park spending. This means they have the power and the incentive to create pork barrel projects. By directing the agency to carry out expensive, unnecessary park initiatives in their own districts, politicians are able to promote local economic development and more jobs. But they also drain funds from the more mundane but necessary maintenance activities at parks such as Yellowstone. A park such as Yellowstone may eventually garner enough attention from Congress through a disgruntled public, but in the meantime natural resources are threatened. Moreover, when money is appropriated, there is an incentive to spend it on high visibility projects instead of behind-the-scenes projects that actually save the environment. Park Service officials would rather attend ribbon-cutting ceremonies at new visitor centers and campgrounds than at sewage treatment plants.

THE WATER PORK BARREL

Problems in our national forests and parks are more than matched by the havoc wreaked by mammoth federal dam projects in the western United States. The era of construction is all but over, but these projects continue to exact a heavy toll on the environment. One of the most egregious examples involves the forced closing of California's Kesterson Wildlife Refuge, an important stopover for millions of waterfowl. The culprit was an unusually high level of selenium, a naturally-occurring chemical, that is benign at low levels but lethal at high levels. Biologists found that as a result of the selenium poisoning, wild duck eggs often did not hatch, and when they did, grotesque deformities were common.

The source of the selenium-laced water that found its way to Kesterson was California's Central Valley Project, an irrigation project that provides subsidized water to farmers in the San Joaquin Valley. Further investigation has revealed that the selenium contamination extends to thousands of evaporation ponds in California's Central Valley and to the rivers that flow into San Francisco Bay.[23]

Another example, one that has taken on tragic proportions, is the precipitous decline in wild salmon stocks in the Pacific Northwest, which salmon supporters contend is due to federal dam projects. As is often the case with these projects, the governmental planning process did not take into account the impact of dams on wild salmon. Initially there were not even fish ladders to allow adult salmon migrating upstream a passage around the massive concrete barriers, so that three dams in the Columbia Basin completely blocked fish passage to historical spawning grounds of some four million salmon and steelhead.[24] And now we understand that even if the fish can get upstream, the smolt have trouble finding their way back to the ocean without assistance from a stream current.

The loss of wild salmon runs on Idaho's Snake River, a major tributary of the Columbia, shows just how bad the situation is. Salmon were the main source of food for the Shoshone-Bannock and Nez Perce Indians for centuries before white settlement in Idaho. Furthermore, until the 1960s, when dam construction began along the Snake River, "Snake and Clearwater rivers supported a thriving salmon fishery."[25] Since 1962, Snake River chinook, sockeye, and coho have all shown drastic declines and are listed under the Endangered Species Act (ESA). Thus, if these salmon go extinct, "the fishing communities and Indian tribes along the Snake River basin will have lost a valuable resource, regardless of whether salmon remain throughout the Columbia basin."[26]

From the Kesterson Wildlife Refuge to the Snake River, water problems are the by-product of the federal government's water pork barrel. For 80 years, the Bureau of Reclamation (BuRec), the agency responsible for making the "desert bloom like a rose," and the Army Corps of Engineers have spent tens of billions of taxpayers' dollars to bring subsidized water to farmers, public utilities, and selected industries. Through interest-free loans and extended repayment schedules, these beneficiaries pay only a fraction of the cost of storing and delivering water. Irrigators whose runoff ends up in Kesterson, for example, pay less than 10 percent of the cost to store and deliver the water.[27] BuRec's efforts have resulted in engineering marvels such as Glen Canyon Dam and Hoover Dam, but the water that continues to flow so cheaply to the few at the expense borne mostly by taxpayers has created environmental problems of tragic proportions.

Again the fiscal and environmental problems inherent in federal water projects are not the fault of bad managers. They result from an institutional framework that does not discipline federal managers to be either fiscally or environmentally responsible. Moreover, the system builds an iron triangle among politicians, water users, and bureaucracies that is difficult to dismantle. If the discipline of free market environmentalism were at work, massive, subsidized water projects would not be built, higher water prices would encourage efficiency, and polluters would be liable for the damage they produce.

BEHIND CANADA'S COD CURTAIN

Political pressures and bureaucracies being what they are, public resource managers simply do not have the incentives to make decisions that lead to sustainable resource use. Nowhere is this more apparent than in the collapse of Atlantic Canada's groundfish fisheries. For centuries, a rich groundfish fishery dominated by cod, haddock, and other groundfish sustained Canada's east coast. Cod landings alone fluctuated around 500,000 metric tons during the first half of the twentieth century. In the 1960s, landings increased dramatically, thanks to foreign fishing and to an increase in the fishing capacity of Canada's domestic fleet.

Over the next two decades, the Canadian federal government continued to expand the domestic fleet, promoting what it called an "expansionist development philosophy" centered on creating jobs.[28] The government set catch limits above sustainable levels for groundfish stocks in order to satisfy the "economic needs of fishing communities."[29] It encouraged entry into the fishery by providing construction and insurance subsidies, tax breaks, and loan guarantees to fishers, boat owners, and processors. Finally, it maintained a labor force well in excess of what could be sustained in the long run by providing overly generous unemployment insurance benefits relative to fisher incomes. Thus, in 1990, unemployment insurance benefits for self-employed fishers in Atlantic Canada were on average $6,600 per year, while income from fishing was only $8,100."[30]

Scientists were predicting dire consequences for Atlantic Canada's groundfish stocks if those policies continued, but the government paid little heed. A 1982 government task force "suggested that the economic viability of the industry, maximization of employment subject to income constraint, and Canadian harvesting and processing should be priorities."[31] By the beginning of the 1990s, the scientists' predictions had come true. Groundfish stocks had collapsed to numbers never seen before. Spawner biomass of the northern cod stock had gone from a high of 1.6 million tons in 1962 to just 22,000 tons in 1992.[32] The story was repeated on a lesser scale for hake, halibut, redfish, haddock, plaice, and flounder. Forced to respond to the collapse in fish stocks, the government declared a moratorium on fishing for cod and many other groundfish. All told, some 500,000 tons less fish were being caught in the 1990s than in the previous decade. Needless to say, the fishing industry has been devastated, with thousands of workers being thrown out of work. Meanwhile, the government has done little to respond to the basic problem of too many fishers chasing too few fish. As a Price Waterhouse study noted, "Although the income-support programs have been effective at providing emergency income support, they have failed to reduce significantly the number of fishers in the industry."[33]

The collapse of Atlantic Canada groundfish stocks serves to illustrate three problems of politically controlled resources:

1. Politicians tend to choose actions with short-run payoffs. In this case, expanding employment in the fishery served to garner votes for the next election, even though such actions proved to be ecologically, economically, and socially disastrous in the long run.

2. Politics can and does override sound science. An internal government document critical of management of the fishery charged that scientific information was "gruesomely mangled and corrupted to meet political ends."[34] The document went on to accuse fishery managers of "scientific deception, misinformation, and obfuscation."[35]
3. Politicians and governmental decision makers are not held personally accountable for their decisions. Since the 1992 moratorium on fishing, 40,000 fishers and processors have been thrown out of work, but no one in government has been fired, demoted, or even reprimanded.

THE WORLD BANK

Established in 1944 at an international monetary conference in Bretton Woods, New Hampshire, the World Bank's mission is to make loans to poor countries for economic development.[36] As of 1997, the bank and its affiliates had over 10,000 employees and $183 billion in loans outstanding.[37] The bank is actually owned by 180 national governments, also referred to as the bank's shareholders. A subset of these shareholders, industrialized countries such as the United States, provide the capital for the bank to lend to poor countries for large-scale development projects such as irrigation, reclamation, and infrastructure, and for broad economic programs such as agricultural resettlement. As noted in various World Bank documents, the bank selects projects that either cannot attract sufficient private capital or that "the private sector alone has found unworthy of investment."[38] Moreover, while one World Bank affiliate, the International Bank for Reconstruction and Development, makes loans at just below commercial rates, another, the International Development Association, makes heavily subsidized loans with funds from triennial grants from industrialized countries. These are long-term loans (35 to 40 years) at no interest and with only a 0.75 percent annual service charge. The World Bank is able to operate in this manner because its loan capital comes from taxes imposed by industrialized countries on their citizens. This also means that there is very little connection between the ability of the bank to obtain funds for future lending and the economic performance of its loan portfolio.

Evidence in recent years indicates that a growing number of the World Bank's projects are in financial trouble. A 1992 report prepared by a high-ranking World Bank official found over one-third of the World Bank's $140 billion in projects to be failing and that deterioration of the bank's loan portfolio was "steady and pervasive."[39] Among the official's noted observations were the bank's "systematic and growing bias towards excessively optimistic rate of return expectations at appraisal" and an "approval culture" in which "staff perceive appraisals as marketing devices for securing loan approval."[40] "Appraisal," the official observed, was equivalent to "advocacy."[41] The official also found widespread noncompliance with the World Bank's loan conditions. Between 1967 and 1989, for example, borrowers had complied with only 25 percent of the financial covenants for the bank's water supply projects.

A number of World Bank development projects have not only been poor economic performers, but also have been catalysts for environmental destruction and social chaos. One of the more notorious examples is the Northwest Region Development Program, known in Brazil as Polonoroeste. Between 1981 and 1983, the World Bank lent a little over $443 million for the paving of Brazilian national highway 364, a 1,500-kilometer stretch of road that connects "Brazil's populous south-central region with the rain forest wilderness in the northwest"; for the building of "feeder and access roads at the frontier end of the highway"; and for 39 rural settlement centers to attract settlers.[42] The intent was to support settlers in raising tree crops such as cocoa and coffee for export. Once the program was under way, however, the Brazilian colonization agency, INCRA, became overwhelmed by the influx of tens of thousands of settlers. Settlers were unable to obtain land titles, agricultural services, and credit. To survive, settlers were forced to burn once-pristine rain forest and plant beans, rice, and maize for food. Crop failure after a year or two on the poor, exposed soils forced many of the colonists to move back to the populated south-central region. Meanwhile, the program increased the rate of forest destruction in a region the size of Oregon from 1.7 percent in 1978 to 16.1 percent in 1991.[43] In addition, settlers and the indigenous population were ravaged by disease. More than 250,000 people were infected with malaria.[44] Some Indian tribes experienced epidemics of measles and influenza, with infant mortality rates reaching 50 percent.[45] Murders, including the highly publicized assassination of Chico Mendez in late 1988, death threats, and assaults were other by-products of the program, as land conflicts among rubber tappers, Indian tribes, cattle ranchers, and settlers were played out.[46]

Brazil was not the only place where a World Bank project wreaked havoc on the environment and caused social unrest. From 1976 to 1986, the World Bank lent about $500 million for another massive resettlement program known as Indonesia Transmigration. The goal was to resettle millions of poor people from the heavily populated inner islands of Indonesia to the sparsely populated outer islands and to provide support for the settlers in the growing of cacao, coffee, and palm oil for export. The outer islands contained 10 percent of the world's remaining tropical rain forest and were inhabited mainly by indigenous tribes. In total, this World Bank program is credited with "the 'official' resettlement of 2.3 million people, and for catalyzing the resettlement of at least 2 million more 'spontaneous' migrants."[47] Like Brazil's Polonoroeste, promised support for settlers failed to materialize, and many settlers were forced into subsistence agriculture that later proved fruitless due to the poor soil conditions for a few years following forest clearing. In wetland and swamp areas, up to 50 percent of the settlers were forced to abandon their sites.[48] Meanwhile, the program led to widespread destruction of rain forest. According to World Bank documents, between 15,000 and 20,000 square kilometers out of a total area of 40,000 to 50,000 square kilometers of rain forest had been cleared as a result of World Bank sponsorship of Transmigration.[49] Equally dismal was its social record. In 1986, the World Bank's own review of the program indicated that 50 percent of the families on program resettlement sites were living below the poverty level, estimated at the time at $540 per year. Ironically, if given the amount the World

Bank paid to install a family at a resettlement site, $7,000, a household could have lived above the poverty level for at least 13 years.[50]

As these two examples demonstrate, economic development is not merely a matter of alleviating capital shortages with government funds; institutions matter. A growing economy going hand-in-hand with environmental quality requires free markets and a strong system of property rights. Countries where property rights are weakly enforced, where the rule of law cannot be counted on, and where government agencies have little accountability tend to have stagnant economies and very little environmental protection.

CONCLUSION

Why do the U.S. Forest Service, the National Park Service, and the Bureau of Reclamation carry out policies that fall far short of environmental protection and fiscal accountability? Why did the Canadian government destroy the cod fishery, and why does the World Bank subsidize destruction of the developing world's environment? Environmentalists who recognize the problem of subsidized destruction of the environment often answer that the cause is people and therefore call for changes in administration. But such an approach overlooks the fact that environmental travesties go beyond people and party lines. Logging may have declined temporarily in the 1990s, but the bulk of spending on U.S. national forests remains with timber production. Yellowstone National Park is found wanting of modern sewage treatment, but Congress ignores the problem, choosing to spend lavishly on elaborate outhouses or on adding parks of questionable national park status. The building of large-scale federal water projects may have slowed in the 1990s, but the environmental travesties are still being felt in terms of lost salmon runs in the Pacific Northwest, and there is little to suggest that the government will fix the problem in the near future. Ignoring the advice of scientists to reduce the number of fishers, the Canadian government subsidized the expansion and eventual collapse of the Atlantic cod fishery. World Bank loans that led to the destruction of tropical rain forests that began in the 1970s and continued through the 1980s demonstrate that public funding is not the solution to economic growth or environmental protection.

The perverse results described in this chapter occur because of institutional failure. In government agencies, bureaucrats have incentives to provide constituents with the products and services they want at little or no cost to them. Entrepreneurs in this arena are rewarded with larger staffs, more authority, and larger budgets, but they do not face the reality check of profitability. Moreover, to the extent that they have a single constituency, such as irrigators vis-à-vis the Bureau of Reclamation, there is little incentive for the bureaucracy to consider other values. Increasing the farmers' incomes with cheap water becomes the primary goal, even if there are perverse environmental consequences. Where there are multiple constituencies, such as those confronting the Forest Service, the political arena becomes a battleground where the developmental interests are pitted against environmental interests in a zero-sum game. Unfortunately, the process is costly to the groups themselves, to the taxpayer, to the economy, and often to the environment. The path to harmonizing economic growth with environmental integrity must include clearly defined property rights and a strong rule of law protecting those rights.

CHAPTER 6

INSIDE OUR OUTDOOR POLICY

Environmental goods generally and recreational opportunities specifically are income elastic; that is, as income rises, the demand for these goods rises faster. Since World War II, incomes for United States citizens have been rising dramatically, increasing the willingness of Americans to pay more for outdoor opportunities. Total visits to Yellowstone National Park in 1997 were 32 percent higher than in 1986; during the same period, entrance fees rose from $2 to $20 per vehicle.[1] Comparing figures from 1960 and 1996, average expenditures per individual, in real terms, more than doubled for fishing and nearly quadrupled for hunting.[2] As a result, a growing number of farmers and ranchers in the United States have found that hunters are willing to pay for the opportunity to hunt on their land.[3]

Increased ability to pay for outdoor recreation means that there are more opportunities for profit and more recreation-related products. The latest U.S. Fish and Wildlife survey of wildlife-related recreation reveals that recreationists spent nearly $61 billion on outdoor equipment and another $30 billion on travel, food, and lodging in 1996.[4] In response, producers have introduced a host of innovative products for enjoying the outdoors, from handheld, electronic location finders to weather-resistant, featherweight clothing.

Until recently, however, a relatively small amount of private money has been spent on the land, water, and other natural resources necessary to produce outdoor recreation as opposed to the equipment for enjoying it. Exceptions include Kampgrounds of America, founded in 1962 as a response to a growing demand for camping facilities along major highways; hunting and fishing clubs in the East, South, and California that have leased lands for many years; hiking clubs that built trails and huts with private initiative; youth camps that provide facilities for a variety of outdoor activities; and ski slopes and lifts that were built on both private and public lands with private funds.

One reason for the low level of private response to growing recreational and environmental demands is that one-third of the land in the continental United

States is controlled by the federal government and is subject to politics. Politicians interested in obtaining votes give their constituents outputs from these political lands at no or low cost. For most national forests, therefore, below-cost recreation costs millions of dollars each year, often exceeding losses from below-cost timber sales.[5] As a result, the private sector has had to compete with a supplier of recreational and environmental amenities that does not have to face the discipline of a profit and loss statement.[6]

As with any good, low or zero fees for federally controlled resources increase the demand and result in overcrowding and diminished quality. In the West, where federal land is pervasive, a growing number of recreationists are finding that the quality of their recreational experience is decreasing. This is causing them to turn their attention to the private sector.[7] Unfortunately, the extent of public landownership combined with the federal government's long history of providing recreational opportunities has created an inertia that is difficult to overcome.

In 1962, John F. Kennedy appointed the first presidential commission on outdoor recreation, an early example of the public's role in providing recreational opportunities. The principal recommendations of the Outdoor Recreation Resources Review Commission emphasized a dominant role for the federal government and led to the creation of several federal programs, including the National Wilderness Preservation System, which has 104,584,556 acres of wilderness lands; the National Wild and Scenic Rivers System, which has 155 rivers totaling 10,902 miles; and the National Trails System, which has 37,357 miles of trails.[8] A subsequent President's Commission on Americans Outdoors in 1987 carried on with this inertia. Charged by President Ronald Reagan in 1985 with reviewing "outdoor recreation policies, programs and opportunities" for the public and private sectors, the commission focused primarily on a single component of recreation: the federal "outdoor estate." Some proposals in the commission's report called for a vast expansion of public landownership and federal controls. These included a $15 billion trust fund that would generate "an absolute minimum" of $1 billion per year to acquire, develop, and protect open space; a nationwide network of public greenways connecting existing and new parks, forests, and other open spaces; and a scenic byways project that would use restrictive zoning and require $200 million per year to protect scenic viewsheds along roadways. In 1987, the president of the Conservation Foundation, William K. Reilly, judged that the commission had "affirmed a crucial federal role in funding, leadership, and resource husbandry."[9]

The recommendation that the government take an even larger role in supplying recreational and environmental amenities assumes that the private sector is unable to provide the optimal amount of recreational services and resource husbandry and that government is better able to do so. A dismayed Jacqueline Schafer, former member of the President's Council on Environmental Quality, pointed out that the Commission on Americans Outdoors did not emphasize people and the ways they create opportunities. She interpreted the commission's report as saying "you can't have [recreational] opportunity unless you have land guaranteed by the government."[10]

While the commission's specific proposals for expanding the government's role have yet to be carried out, the push for more public landownership and control continues in the political arena. On January 11, 1999, President Clinton proposed his "Lands Legacy" initiative, which would allocate a onetime $1 billion for protecting wilderness, parks, and coastal areas. Nearly half this amount would be used for land acquisition, including 450,000 acres near and in Mojave and Joshua Tree national parks in California; land inside Florida's Everglades National Park; 100,000 acres in national forests and wildlife refuges in New England; and parcels along the Lewis and Clark Trail and at Gettysburg, Antietam, and other Civil War battlefields.[11] The announcement came on the heels of Vice President Al Gore's "smart growth" initiative, which would use $1 billion in tax credits and federal grants to help communities put the brakes on private land development. In particular, the federal government would provide support to "communities that work together on smart-growth plans and allow more federal transportation dollars to be used to reduce gridlock, expand mass transit and encourage regional planning."[12] Of course, this presumes that urban sprawl is rampant in America and that, therefore, the federal government needs to expand its role here as well.[13]

Ignored in these proposals is how owners of critical private lands are responding to the increasing demand by providing recreational opportunities and environmental amenities, particularly as they relate to wildlife. Private lands "constitute 60 percent of the 1.35 billion acres of America's forests and rangelands," and they provide some of the best habitat for game and nongame wildlife in the United States.[14] Also ignored is the federal government's record of stewardship, which has not always been good. With thousands of acres added to the federal estate each year, operating budgets are stretched. This, combined with poor incentives for good stewardship, raises questions about the federal government's ability to meet the demands for recreation and environmental amenities.

THE ECONOMICS OF OUTDOOR RECREATION

To understand how the private sector can provide more outdoor recreation, it is necessary to recognize that individuals respond to prices. Consumers move away from buying relatively high-priced goods by finding lower-cost substitutes. If recreational opportunities are available at a low or zero price, then consumers can be expected to take advantage of those opportunities and use them to the point at which the additional value in consumption is equal to the additional cost. At the same time, producers will not shift resources away from alternative productive uses, such as farming or ranching, and into recreation when prices for such activities are not high enough to yield positive returns.

Product substitution is often ignored in the formulation of natural resource policy because prices faced by decision makers are often zero or nominal. For example, the Gallatin National Forest in southwestern Montana attracts thousands of fishers, hunters, hikers, and campers every year. Yet, the Forest Service has failed to monetize the value that recreationists place on the area. The fees charged are so low that revenue from recreation amounted to $410,000 in fiscal year 1996, and the costs of provision, including trail and road construction and maintenance, campgrounds, and administration, totaled $2,100,000.[15] At these low prices, there is no signal that recreational resources are scarce and little incentive for consumers to consider substitutes.

The same set of conditions holds for producers. A rancher who owns land along a trout stream could significantly improve fishing by keeping cattle away from the stream, thus reducing bank erosion and increasing bank vegetation. The capital cost of fencing out the cattle may be quite low, but if the rancher is prevented from charging anglers an access fee and making a profit, he has little incentive to put up the fence. Improving fishing opportunities would only encourage fishers to enter the rancher's property, reducing his privacy and exposing him to potential liability.[16]

Environmentalists tend to argue that there is no (or at least very little) substitution between traditional commodities produced from land and water and environmental amenities. For example, environmentalists have fought for two decades to prevent exploratory oil drilling in the 1.5-million-acre coastal plain of Alaska's Arctic National Wildlife Refuge, an area valued for its environmental amenities and "potentially enormous oil and gas resources."[17] On this issue, however, environmentalists have given little room for compromise: "Where's the compromise? It's the type of an issue where you don't think compromise. You drive a stake in the sand and say, 'You don't cross the Canning River [the western edge of the refuge's coastal plain].' "[18] In 1995, both the U.S. House of Representatives and the U.S. Senate passed legislation to open the coastal plain of the Arctic National Wildlife Refuge to exploration, but the legislation never became law. The budget bill containing the legislation was vetoed by President Clinton.

Yet, in some cases, traditional outputs and environmental amenities can be jointly produced with only a small disturbance to the environment. In fact, certain arrangements can be mutually beneficial to both environmental and developmental interests. On the Audubon Society's privately owned Rainey Preserve in Louisiana, for example, petroleum development operates under carefully controlled conditions in an environmentally sensitive sanctuary for wildlife. At Rainey, developers produce gas from a large reserve, and owners of the sanctuary receive royalties from the activity, providing them with additional resources to purchase other sensitive lands.

In other cases, amenities and recreation can be jointly produced with traditional commodities using the same inputs. In these cases, a rise or decline in the supply of traditional commodities will result in a corresponding rise or decline in the supply of amenities. Agriculture and open space are examples of jointly produced outputs. About 1.5 billion acres of crop-, forest-, and pasture- or range-lands provide a substantial portion of open space in the United States.[19]

Recreation and timber can also be jointly produced. Growing timber helps provide natural water storage for streams and rivers and ensures stable flows throughout warmer months. These conditions support the burgeoning demand for water-related activities, such as fishing and white-water rafting, as well as pleasing settings for camping and hiking. These areas also provide valuable habitat for big game species, such as elk, deer, and bighorn sheep, and important habitat for endangered and threatened species, including grizzly bears and peregrine falcons. Timber removal, however, can reduce the recreational amenities if it is not managed to take account of the impact on watersheds, riparian zones, and wildlife habitat.

IS THERE MARKET FAILURE?

From the standpoint of supplying both traditional commodities and outdoor recreation, the important public policy question is whether markets accurately reflect recreational and environmental values (or costs) to consumers and producers. The answer to this question depends on the nature of the property rights to inputs and outputs or the ability of input owners to contract with one another and with consumers.

Cattle ranching provides one example of how property rights enable markets to work effectively to provide meat. Combining consumer willingness to pay a price that makes beef production profitable with well-defined and enforced property rights ensures that producers can capture the profits from cattle production. In this way, the enabling mechanisms for property rights include: (1) fencing, which defines the owner's land and grass used in production; (2) a brand, which identifies ownership of the cattle; and (3) a legal and political system, which enforces the owner's claim to the inputs and the outputs.

Consider how these same elements apply to the private provision of wildlife and wildlife habitat. Economist Harold Demsetz used the example of the Montagnais Indians in the Labrador Peninsula, who established beaver trapping territories during the early 1600s, to show how these property rights can work.[20] Before white trappers and traders arrived in the region, the Montagnais hunted beaver communally. But as the demand from new markets grew, the value of the beaver increased and more pressure was put on the resource. To avoid complete depletion of the beaver population, the Montagnais established private hunting grounds and successfully managed the beaver on a sustained-yield basis. The costs of defining and enforcing the property rights were the time and resources used to identify territories for individual hunters and to exclude others from those territories. When the value of the beaver was high enough, it was worth the cost of establishing rights to the animals.

Property rights are equally crucial today in the private provision of wildlife habitat. For a farmer, this provision of habitat can entail leaving a marsh for nesting waterfowl instead of converting it to cropland. For a rancher, it can entail establishing nongrazing riparian zones to protect streamside habitat. But producing a marketable product requires owning enough habitat to account for the migratory nature of wildlife and being able to charge demanders of the habitat for their enjoyment of it. In the case of controlling enough habitat, obviously the land requirements will vary with the species in question. For ground squirrels or prairie dogs, the land requirements will not be nearly as great as they will be for a migratory elk herd. The cost of amassing the necessary land, of course, will depend on how fragmented the ownership is and how difficult it is to contract for attributes of the land necessary for the species in question. For example, it will be much more difficult to put together enough land for wildlife habitat in a suburban area with one-acre lots than in a rural area with 10,000-acre ranches. If the contracting obstacles for putting together the necessary land and for charging the users can be overcome, owners can claim the rewards of producing wildlife and wildlife habitat. Whether it is worth overcoming these obstacles will depend on what consumers are willing to pay. Hence, if consumers are not willing to pay much, if land requirements are large relative to the fragmented ownership patterns, or if trespass is difficult to prevent, then there will be little private provision of the environmental good.[21]

The contracting problems are further exacerbated by wildlife laws that establish state ownership of the wildlife and do not allow wildlife agencies to contract with landowners to produce wildlife and wildlife habitat. Under such laws, landowners control access to their property, but the state regulates the taking of wildlife on the property. This bifurcation of control can prevent private provision of wildlife.

On the other hand, if laws allow contracting between the state wildlife owners and landowners, private provision becomes much more feasible. Suppose extended hunting season on a ranch would be possible if the rancher improved habitat. The extra days of hunting would allow the rancher to collect additional revenue and capitalize on improvements.

The historic reason offered for state control of wildlife is that markets were responsible for the decimation of populations in the late nineteenth and early twentieth centuries. Proponents of state ownership argue that in the absence of government ownership, commercialism of the bison almost led to its extinction. The problem, however, was not private ownership of bison, but rather, no ownership of bison until they were killed and in possession of the hunter. It was a classic tragedy of the commons and certainly not a failure of the market.[22]

The same misguided view of markets for wildlife led to the abolishment of the legal ivory market by the 1989 CITES treaty.[23] Western conservation and animal rights groups that pushed for the ban interpreted the legal market for ivory as the culprit in the decline of African elephants due to poaching. While the ban was successful in stopping the legal trade in ivory, which resulted in lost revenues for local conservation efforts in African countries, there is little evidence that it stopped poaching. Poaching, in fact, has "continued without interruption in Africa's tropical forests."[24] In Zambia, Tanzania, Cameroon, Ivory Coast, and Zaire, poaching has slowed, but that has been due to increased law enforcement financed, in part, by outside donations.[25] The Kenyan government does attribute a 32 percent increase in its elephants between 1989 and 1991 to the ban, but it is more likely due to elephant immigration from adjacent countries or inaccurate population estimates.[26] In Zimbabwe and Botswana, elephant populations have thrived for years because of successful conservation programs that capitalized on the legal ivory trade. In those countries, local people have a strong economic interest in protecting elephants from poaching because the revenue from tusks and hides and a portion of the money made from selling hunting permits go to local communities where the elephants live. It is property rights and markets that account for the rapid elephant population growth in Zimbabwe and Botswana and for the declining population in the rest of Africa, as shown in Figure 6.1.

Similarly, when landowners in the United States are compensated for the wildlife on their property, the fate of wildlife and wildlife habitat can change dramatically. Montana rancher Franklin Grosfield was tired of losing hay to wildlife and of being awakened in the middle of the night by hunters seeking permission to hunt on his land. But attitude toward these costs changed when he decided to lease his land to a hunting club. The hunters provide Grosfield with revenue to supplement his cattle operation, and, as he put it, "I've taken one of our worst liabilities [wildlife] and turned it into an asset."[27]

If ranchers cannot capture benefits from producing environmental amenities, those amenities may become liabilities. Rancher Michael Curran identified one of the factors that influence landowners' decisions to produce wildlife and habitat: "We feed 250 elk for six months, and 500 deer and about 300 antelope for an entire year. . . . We've figured that if the Montana Fish and Game Department paid us for the forage consumed, they'd owe us $6,000 every year."[28] With no reward for harboring wildlife, the presence of deer, antelope, and elk on private land is a liability rather than an asset, just the opposite of what Franklin Grosfield achieved.

Even in the public sector, returns from wildlife and other amenities relative to traditional commodity uses affect resource spending decisions. The Forest Service does receive revenue from logging and grazing, but on most national forests it does not charge for activities such as hunting, fishing, bird watching,

Figure 6.1
Percent Change in Elephant Populations, 1989–1995

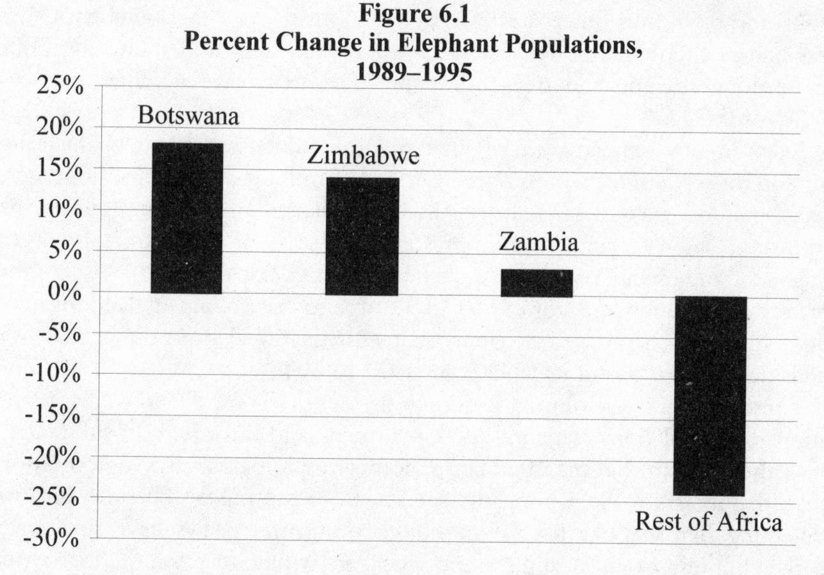

Source: Getz et al. *Science,* March 19, 1999, 1855.

hiking, and nature trips.[29] As a result, for fiscal years 1994 through 1996, the Forest Service derived 84 percent of its annual total revenue from timber sales, while it derived only 6 percent from outdoor recreation. Over the same period, it spent 34 percent of its budget preparing and administering timber sales and only 12 percent of its budget on wildlife and fisheries, soil, water, air, ecosystem planning, and watershed restoration.[30]

By providing outdoor recreation for free (or nearly so), public land agencies not only subsidize recreational activities and hence give them short shrift in management decisions, they make it difficult for adjacent private landowners to compete with the free good. Timber companies with large landholdings in the West could provide significant recreational opportunities, but they cannot compete with the zero price the government has set for activities such as fishing and hunting on adjacent national forests. As a result, timber companies in this region spend little to enforce their property rights in amenities, and they tend to ignore these values in resource decisions. This is in sharp contrast with the East and the South, where there are much fewer acres for federally subsidized recreation.

HUNTING FOR HABITAT

In the East and the South, private landowners manage for land values other than just timber, and they do so at a profit. The International Paper Company's wildlife and recreation program is a prime example. International Paper employs specialists to oversee wildlife and recreation on its lands, including the 16,000-acre Southlands Experiment Forest located near Bainbridge, Georgia. At Southlands, researchers develop forest management practices that enhance wildlife populations as well as profits. White-tailed deer, turkeys, rabbits, bobwhite quail, mourning dove, and other species are beginning to reap the benefits of these new management techniques. Habitat is improved by controlled burning, buffer zones along streams, and tree-cutting practices that leave wildlife cover and plenty of forage.[31]

According to company officials, investing in wildlife research and habitat production makes sound business sense. On its 1.2-million-acre mid-South region, which includes parts of Texas, Louisiana, and Arkansas, profits from hunting, hiking, fishing, and camping are an impressive 25 percent of total profits. By far the biggest revenue generator is the multiyear hunting lease. In early 1994, approximately 2,100 clubs paid from $2 to $5 per acre to lease company land for hunting.[32] Figure 6.2 shows what happened to International Paper's revenues from recreational activities on its lands from 1977 to 1998.

These returns from hunting leases pay dividends for wildlife as well as for the company. Populations of deer, turkey, fox, quail, and ducks are up substantially since the program began. In addition, company biologists carry out an assortment of projects to help nongame species, from putting up bluebird boxes to protecting heron rookeries. Even though nongame species have no explicit market, hunters, campers, anglers, and hikers are willing to pay more for a diversified experience.

In another corner of the United States, North Maine Woods, Inc., offers

Figure 6.2
International Paper's Recreational Revenues for Mid-South Region

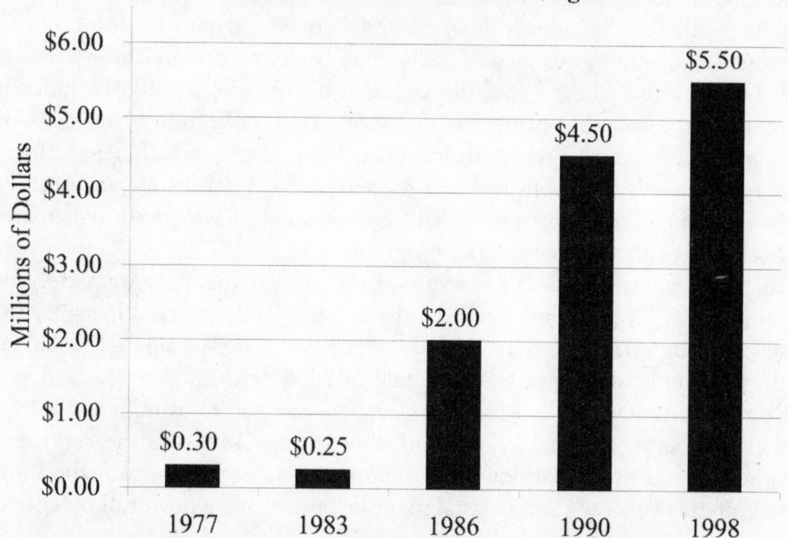

Source: Richard Boitnott of International Paper Company and Tom Bourland of Crawford and Bourland Consulting Foresters, 1999.

another interesting contrast to public land management. A nonprofit association formed by 20 landowners, North Maine Woods manages recreation on just over 2.8 million acres (about twice the size of Delaware) of mostly private commercial forests. The area includes two of the wildest rivers in New England, the Upper St. John and the Allagash, both of which have numerous stretches of white water desirable for canoeing. The area also has abundant wildlife, including moose, white-tailed deer, black bear, and partridge. With 252 lakes and ponds and miles of brooks and streams, the area is noted particularly for its excellent fishing for brook trout, lake trout, landlocked salmon, and whitefish. Most of the area is not considered a wilderness because it is managed for timber production and is interlaced with logging roads, but it still provides a high quality outdoor experience. Between 1977 and 1997, visitor days in the area grew from 120,000 to over 200,000.[33] This positive trend was accomplished without any advertising by North Maine Woods, Inc., and is thought to be due in part to recreational users spreading the word about their favorable experience visiting the area.

Landowners formed North Maine Woods, Inc., when they began to experience problems from recreational use. They were suffering the costs of erosion and safety on private roads, overcrowding and overuse of camping areas, littering, and the ever-present threat of forest fires. The association's primary goal was to develop a program to manage public use on their lands and to find ways to fund it. North Maine Woods, Inc., now controls access to the area through 17 checkpoints and access roads, where visitors are required to register, pay fees for different types of use, and obtain permits for campsites. In 1998, the fees for Maine residents ranged from $4 a day to $60 for all-season use with camping privileges; the fees for nonresidents ranged from $7 a day to $100 for all-season permits with camping privileges. Individuals under fifteen and over seventy are not charged a fee. Since 1986, the association's program, which includes constructing and improving campsites, running a trash collection system, and running public education programs, has been financially self-sufficient. Although the organization's initial efforts were resisted by those who were accustomed to free, unrestricted access, recreational improvements have promoted a greater appreciation of the benefits of the program.[34]

In Texas, where about 98 percent of the land is privately owned, hunting leases and habitat are being provided by smaller landowners, with areas under lease averaging 367 acres. A 1996 study of hunting leases with ranches and farms in the state reported an average lease rate of $634 per hunter per season, with a price range from as low as $100 per hunter per season to as high as $4,250 per hunter per season, depending on the quality and quantity of game, services, and facilities offered by the landowner. Moreover, the net return from the hunting lease "often far exceeds agricultural income," providing a powerful incentive for landowners to provide quality hunting opportunities.[35]

Since many outdoor amenity markets are still developing, a number of enterprises have emerged to bring potential buyers and sellers together. Based in northern California, Multiple Use Managers, Inc., for example, offers expertise in wildlife management to landowners who want to develop quality hunting opportunities on their properties and markets those opportunities to hunters by booking hunts and supplying guides, accommodations, and food.[36] Based in southwestern Montana, Fay Fly Fishing Properties, Inc., advertises in various fly-fishing publications nationwide for the purpose of attracting "conservation-minded fly fishers to purchase land along Montana's rivers and streams."[37] Serving as a broker for the buyer, the company researches both unlisted properties and properties listed by real estate firms to find the property that best fits the buyer's needs. The company also "works extensively with conservation easements" to help farmers and ranchers keep their traditional agricultural businesses intact.[38]

As Aldo Leopold recognized decades ago, the recreation market benefits the sportsman and the environment as much as it does the landowner. He argued that there was no need to reject market forces simply because "such tools are impure and unholy."[39] He understood the importance of incentives and asked whether there was any way "to induce the average farmer to leave birds some food and cover without paying him for it." The *Fishing and Hunting News* nicely summarizes the benefits to the consumer of compensating landowners for providing wildlife habitat and recreational opportunities:

> In these days of posted farmland, shrinking public access, and growing hordes of hunters, a hunting preserve membership is an absolute guaran-

tee that you will have a place to hunt and a place to take junior, and you won't have to spend half of the day looking for a landowner whose permission to hunt may not come readily. . . .The bottom line is better hunting, more shooting, and a happier end to each excursion. What more can the sportsman ask for?[40]

BEYOND HUNTING AND FISHING

Private land conservation efforts are another way in which markets are helping protect environmental amenities. Land conservation trusts with tax-exempt status are flourishing around the country for the purpose of preserving land for its environmental values and for keeping land in agricultural uses. Funds are raised by soliciting members, who pay a small fee each year, and by soliciting much larger grants from foundations and corporations. Land trusts use these funds to buy land in fee simple title, to purchase conservation easements, or to protect donated easements. In 1950, only 36 conservation organizations existed in the United States; by 1975, there were 173; by 1982 there were 404 groups representing over 250,000 members; and by 1998, there were 1,213 local conservation organizations controlling 5 million acres throughout the United States, Puerto Rico, and the Virgin Islands—more than twice the amount reported in 1988.[41]

In addition to membership fees and philanthropic grants, land conservation organizations often rely on user fees to help make ends meet. Speaking for the Trustees of Reservations in Massachusetts, Gordon Abbott, Jr., stated: "We're . . . fortunate that user demand enables us to raise 35 percent of our operating income from admission fees and that these can be adjusted within reason to catch up with inflation. We're great believers in the fairness of users paying their way."[42]

At the national and international levels, the Nature Conservancy leads the way in private land conservation. As of June 1999, it owned approximately 4,192,000 acres of land in the United States and 425,000 acres in Latin America and the Caribbean.[43] The conservancy is also a pacesetter in finding ways to raise money to cover operating expenses of its preserves. On the 13,000-acre Pine Butte Preserve in northwestern Montana, for example, the conservancy offers nature tours through the last prairie grizzly bear habitat in the contiguous 48 states. It oversees cattle grazing on selected areas of the preserve, where grazing fees netted $10,000 in revenue in 1986. The conservancy also started a guest ranch business, offering guided nature tours access to hiking trails, fishing, and horseback riding. The revenues from the ranch help offset the cost of operating the preserve.[44]

GETTING IN THE WAY OF MARKETS

The success of the private recreational and conservation sector described above has come despite competition from subsidized provisions of environmental amenities on federal lands. This explains why International Paper has an active hunting and recreational lease program in the mid-South region, where most of the land is privately owned, but has no recreational program in the Pacific Northwest, where national forests provide free access. The price distortion resulting from subsidized recreation on federal lands has thwarted private recreational alternatives.

The free recreational alternative, however, is partially disappearing on some federal lands as a result of the Recreational Fee Demonstration Program implemented in 1996. This program allows the National Park Service, the Bureau of Land Management (BLM), the U.S. Fish and Wildlife Service, and the Forest Service to raise user fees at 400 demonstration sites in the federal land system. Demonstration sites are allowed to keep at least 80 percent of the proceeds for funding "a broad array of activities ranging from costs of fee collection to resource preservation and law enforcement."[45] In fiscal 1996, the four participating agencies collectively took in $84.4 million from the program. These same agencies are projected to take in $186 million from the program in fiscal 2001.[46] The program is not only helping modernize fee collection, it is contributing to repair and maintenance and customer service. For example, the Park Service used fees to rebuild 5,000 feet of deteriorating trails at Natural Bridges Monument in Utah, and the Forest Service used fees to keep open three visitor centers at Mount St. Helens National Monument in Washington. Though most fees are relatively modest, the BLM and Forest Service have begun experimenting with more aggressive pricing schemes at some popular sites in an attempt to reduce congestion during weekends and holidays.[47] It remains to be seen whether these agencies will implement more realistic prices for high demand, upscale recreational activities such as trophy elk hunts in the backcountry of public lands. Hunters pay outfitters anywhere from $3,500 to $4,000 to take them on such hunts but pay no hunting fees to those who control the habitat.[48]

Government subsidies for traditional commodity outputs also distort prices and hinder the private provision of environmental amenities. Federal farm programs, for example, have paid farmers to grow crops such as wheat, corn, cotton, and rice, encouraging them to convert wetlands, woodlands, and grasslands to croplands.[49] Populations of ducks, pheasants, quail, and cottontail rabbits have drastically declined over the last 40 years because their habitat has been lost to subsidized crops. Ironically, under pure market conditions, farmers would not be able to farm areas of poor quality and make a profit, but the subsidies provided by the federal government have enabled them to ignore economic reality.

Legal restrictions on private property rights provide another example of governmental impediment to the private provision of recreational and environmental amenities. Consider the provision of stream habitat for fishing. As we will see in chapter 8, some riparian landowners are taking actions to improve habitat and charge fees for fishing access, but there are impediments to this market created by the public trust doctrine. Evolving from English common law, which prevented the Crown from excluding citizens from using navigable waterways, tidal areas, and beaches, the public trust doctrine holds that rights to water and riparian land are subject to the state's trust responsibility to protect resources. California's famous *Mono Lake* case in 1983, for example, forced the state to order Los Angeles to stop dewatering the lake because it was destroying the aquatic habitat held in trust by the state.

This doctrine was greatly expanded in 1984 when the Montana Supreme Court ruled that the state held all waters in trust for the people and therefore that the people could not be denied access to their water even if it flowed across private land. The ruling was codified by the state legislature in 1985 to limit access to public access points such as where streams flow under highways and between the high watermarks.[50] As a result, it is more difficult for landowners to control access and to benefit from riparian habitat provision. For example, in February 1999, the state Fish and Wildlife and Parks Commission ruled that property owners Charles and Elena d'Autremont had no legal basis for limiting access to a one-and-one-half-mile stretch of the Ruby River that courses through their land in southwestern Montana. The couple argued that the river was suffering from overuse by anglers. They wanted to limit access to protect the river and its fishery, but public-access groups protested on the grounds that it violated the stream access law.[51]

A similar barrier results from state ownership of wildlife. With wildlife the property of the state, it cannot be sold or regulated for hunting by anyone other than the state. This legal formality is partially circumvented by trespass and access fees charged by landowners, but some states explicitly restrict the ability of the landowner to sell permits. Indeed, some wildlife groups argue that the public trust doctrine should apply to wildlife as well as water, and therefore people should not be denied access to their wildlife. Although free access to wildlife is not guaranteed by the public trust doctrine, some wildlife organizations argue that a publicly owned resource such as wildlife should never be subject to access or user fees.[52] If this were to happen, the potential for landowners to profit from providing wildlife habitat would be greatly curtailed and hence the incentive removed.

In contrast to the public trust doctrine approach to wildlife, Colorado, California, and Utah have chosen to harness positive incentives through their "ranching-for-wildlife" programs. Under these programs, the landowner might agree to improve brush cover for upland game birds or plant willows to provide habitat for white-tailed deer. In return, the state allows a modification of hunting regulations on the enrolled ranch properties so that landowners can raise additional revenues from wildlife production. This typically includes extending the hunting season, modifying the limit on game taken, or selling permits directly to hunters without going through the state lottery system.[53]

Another legal barrier to private provision of riparian and aquatic habitat results from the "use it or lose it" principle in western water law. As a way of validating water rights, states apply the rule of beneficial use, which states that users who withdraw water from the stream must put it to beneficial use. This generally means that beneficial private use requires diversions and therefore excludes instream flows. In other words, if a right holder decides not to divert all the water to which he is entitled and instead leaves some of it in the stream for fish, he loses his claim to it because that use is not considered beneficial. In 1917, for example, a Utah court found it

> utterly inconceivable that a valid appropriation of water can be made under the laws of this state, when the beneficial use of which, after the appropriation is made, will belong equally to every human being who seeks to enjoy it. . . .[We] are decidedly of the opinion that the beneficial use contemplated in making the appropriation must be one that inures to the exclusive benefit of the appropriator and subject to his dominion and control.[54]

The result of this "use it or lose it" principle was manifested on Montana's Ruby River in May 1987. Minimal snowpack, little spring rain, and heavy demand for irrigation reduced a 1.5-mile section of the Ruby to a trickle. Hundreds of trout were stranded in overheated and deoxygenated pools and eventually died.[55] The river could have been kept flowing and the trout kept alive if only small amounts of water had been transferred from irrigation to instream flows. The amount of water necessary to have prevented the kill could have been rented for less than $4,000. With 50,000 members and an annual budget of over several million dollars, Trout Unlimited had access to the necessary resources to purchase the water, had it been permitted to do so.[56] Because the "use it or lose it" principle would not allow such a transaction, the only option, short of contentious litigation that would have taken too long anyway, was for the state to rely on the good graces of water owners. Montana's Department of Natural Resources and Conservation did eventually persuade local irrigators to leave approximately a hundred cubic feet per second flowing in the stream, but the effort was too little and too late for the trout.

Government land use regulations are another source of disincentives for landowners interested in improving wildlife habitat. For example, when Dayton Hyde put 25 percent of his Oregon ranch into marshes for wildlife, initiated research on the sandhill crane, and built a lake with three and a half miles of shoreline for wildlife, his use of the land was curtailed because it might have affected the species he had attracted with his improvements. He paid a price for his good deeds: "My lands have been zoned. I am being regulated for wetlands that weren't there before I created them. Like most of my neighbors I can save myself from financial disaster only by some creative land management, but the state legislature has cut out most of my options."[57] Despite this penalty, Hyde has continued to promote private conservation through Operation Stronghold, an international organization he founded to help landowners with conservation efforts. However, because of experiences such as his, Hyde has found landowners reluctant to join. As one landowner put it, "Look, you don't understand. We would like to do our share for wildlife but we are afraid if we create something worthwhile the public will want what we have. It's just plain easier and a lot safer to sterilize the land."[58]

The Endangered Species Act (ESA) provides another example of perverse incentives. In an effort to protect endangered species, the ESA makes it illegal to "take" a listed species, meaning "to harass, harm, pursue, hunt, shoot, wound, kill, trap, capture, or collect, or to attempt to engage in any such conduct."[59] In other words, if a landowner's actions are interpreted as a "take," land uses may be strictly regulated. This might encourage landowners to "shoot, shovel, and shut up."

But landowners do not have to go this far; they can take perfectly legal preemptive action to keep the species off private property. A famous North Carolina case shows how this works with the endangered red-cockaded woodpecker (RCW), which lives in old-growth pines. After Ben Cone was prevented from harvesting 1,500 acres of his 7,200-acre property because it was home to RCWs, he started cutting his trees at forty years of age instead of eighty, thus eliminating the old-growth trees in which RCWs might live.[60]

Supporters of the ESA contend that such cases are isolated acts carried out by lawbreakers, but a study by Dean Lueck and Jeffrey Michael shows systematic evidence to the contrary.[61] Examining hundreds of logging operations in North Carolina, Lueck and Michael find that the average age of harvest falls significantly as the prospect of having RCWs occupy a site increases. Table 6.1 shows that the average age of harvest falls from nearly sixty years if there are no RCW colonies nearby to thirty-six if there are 25 colonies within 25 miles of the logging site, and to sixteen years if there are 437 colonies (the densest populations in the state) within 25 miles. They conclude that their "finding validates the concerns of some environmentalists who have noted that RCW populations have been declining on private land during the 28 years the red-cockaded woodpecker has been regulated by the ESA."[62]

WHERE TO FROM HERE?

The first presidential commission on outdoor recreation in 1962 helped promulgate the myth that more federal lands and more regulation of private lands and water are necessary to meet growing recreational demands. Partly as a result of this mentality, more than thirty million acres, an area the size of Florida, have been added to the federal estate since 1960, and countless regulations have been imposed on private landowners in the name of habitat conservation. The second presidential commission in 1987 continued this tack, calling for more fed-

Table 6.1 Predicted Harvest Age by Ownership Type and the Number of RCW Colonies Within a 25-Mile Radius. FIA Data.

Number of RCW colonies within 25-mile radius	Industry-owned land	Nonindustrial private land	All privately owned land
0 colonies	58.11 years	54.34 years	55.26 years
25 colonies	35.97 years	47.95 years	46.68 years
437 colonies	16.30 years	42.42 years	39.07 years
Average age at harvest	39.4 years	50.20 years	47.90 years

Source: Dean Lueck and Jeffrey Michael, "Preemptive Habitat Destruction Under the Endangered Species Act," unpublished manuscript, Department of Agricultural Economics and Economics, Montana State University, Bozeman, MT, 1999.

eral land acquisitions and more restrictions on private landowners. In 1999, the Clinton administration called for the expenditure of $1 billion per year from the Land and Water Conservation Fund to purchase even more lands for environmental purposes.

This continuing expansion of the federal government's role in outdoor recreation and conservation ignores the important role that the private sector is playing in responding to consumers. From for-profit firms such as International Paper to not-for-profit groups such as the Nature Conservancy, entrepreneurs are finding innovative ways of conserving land and water.

There are several policy reforms that could further encourage the private sector in its recreation and conservation endeavors. The least politically palatable, but perhaps the most effective, would be to privatize or lease some federal lands and let the private sector manage them for recreational and environmental amenities.[63] Short of privatization, state agencies and courts could get out of the way of establishing private rights to wildlife, instream flows, and riparian habitat so that markets for these ecosystem services can form. Instead of promoting a public trust approach that opens access to water, land, and wildlife, policies can encourage contracting for access so that recreational and environmental resources become assets, not liabilities. The Endangered Species Act and other land use regulations should be reconsidered in terms of the perverse incentives they create for landowners to take preemptive action to eliminate wildlife habitat. And the federal government should continue moving in the direction of eliminating recreational subsidies by charging realistic user fees for federal lands. Not only would this be more fiscally responsible, it would make it easier for the private sector to compete with the public sector. Rising values of recreational and environmental amenities will provide an incentive for entrepreneurs to develop new contractual arrangements for overcoming the free-rider problems associated with many environmental amenities, but getting the government out of the way will also help.

CHAPTER 7

ECOLOGY AND ENERGY:

Prospecting For Harmony

When energy resources are controlled by the political sector, environmentalists are pitted against exploration and development companies in a setting that encourages confrontation. This problem became especially acute during the 1970s, when the energy crisis focused a great deal of attention on the overthrust belt in the Rocky Mountains, where the federal government controls nearly half of the land. Since then, the controversy has waxed and waned with fluctuations in energy prices.

One place where environmentalists continually battle energy producers is the Arctic National Wildlife Refuge (ANWR), a 1.5-million-acre coastal plain on the north slope of Alaska. According to the latest assessment, the U.S. Geological Survey raised the oil potential of the coastal plain to between 4.3 and 11.3 billion barrels of recoverable oil, with a mean of 7.7 billion barrels.[1] The area is also important for wildlife, including 500 musk oxen, 100 Alaskan brown bears, dozens of wolves, 325,000 geese, 300,000 other wildfowl, 62 marine species, and, for brief periods during the summer, more than 100,000 caribou of the Porcupine herd. In response to recommendations that the area be made available for oil and gas development, environmentalists have mounted an effective campaign to block final approval.

In debates such as that over ANWR, environmentalists contend that energy exploration and development inevitably lead to environmental damage, but this need not be the case. In fact, there are approaches that enable development to be carried out in a sensitive manner. For example, under certain conditions developers can use directional or slant drilling to prevent disturbance to fragile surface areas such as swamps and marshlands. This approach has been used successfully on the Michigan Audubon Society's Bernard W. Baker Sanctuary. Special applications, such as highly insulated platforms, have also been used to mitigate disturbances to fragile Arctic environments. Continued field research in the tundra region of Alaska's North Slope, the coastal marshes of Louisiana, and the steep mountainous areas of the Rockies has led to new ways of minimizing the effects of oil and gas activities on wildlife and habitat.

Given that these techniques have been tried and proven effective on lands controlled by the environmental groups protesting exploration and development in ANWR, why do they continue to resist oil and gas exploration? Free market environmentalism says the answer lies in the incentives faced by the parties involved in the political allocation of energy resources.

COSTS ARE SUBJECTIVE

There is no question that oil exploration and production can have adverse effects on the environment. Noise, lights, moving earth, oil spills, and blowouts are only a few of the concerns environmentalists have when exploration and development take place in environmentally sensitive areas. Slant drilling, quiet mufflers, double- or triple-hulled tankers, and interruptions in drilling schedules to accommodate wildlife can reduce the risks of environmental damage, but obtaining these benefits means higher costs for development companies. Thus the question is whether the tradeoffs are worth it and who will make this decision. What are the environmental costs of exploration and production, and how should the risks of incurring higher costs from accidents be weighed?

Environmentalists and developers have such conflicting perspectives that they will see the risks of accidental damage very differently. Developers, striving to keep production costs low, are likely to prefer lower precautionary costs and are willing to take higher risks for accidental damage. At the same time, if they face the full costs of damage, their decisions must weigh the savings of less precaution against the possibility of paying higher costs if damage occurs. Environmentalists prefer a lower probability of environmental damage and, hence, higher production costs. But unless risk abatement represents an actual cost to them, they sacrifice nothing in demanding the lowest possible risk and highest possible cost of production. In political debates such as ANWR, the cost to environmentalists of saying no is zero, and the cost to developers of taking precautions depends on politically determined regulations. Not surprisingly, both sides have little incentive to negotiate.

The free-market-environmentalism question is how can the costs and benefits of energy exploration and development be internalized to the companies that want to maximize profits and to groups that want to protect the environment. The information necessary to assess the appropriate level of environmental damage abatement will vary considerably from location to location and from time to time. Although experts can assess the time-and-place-specific environmental and technological constraints, all costs ultimately are subjective and depend on the values that individuals place on resources. Even if everyone could agree on the probability of environmental damage associated with oil development, there is still the problem of agreeing on what the subjective costs are.

Consider oil exploration on public land near Yellowstone National Park. For one person, a drilling rig might represent nothing but noise and visual pollution that significantly reduces the value of a visit to the park. For another, the close-up view of the massive, lighted derrick drilling 24,000 feet below the Earth's surface may represent a fascinating contrast between impressive technology and beautiful scenery. And for the roughneck whose job depends on oil exploration, pristine mountains or impressive technology may be irrelevant. None of these values is right or wrong; each simply represents a special interest. The problem is getting individuals with diverse preferences to accurately and honestly reveal their values in the political setting. A person who sees the drilling site as a form of pollution will most likely claim that no drilling is acceptable in a pristine mountain setting. The person whose job depends on oil exploration may contend that a pristine environment is of little value if he is unemployed or if gasoline is not available to transport people to the woods.

POSITIVE- OR NEGATIVE-SUM GAMES

In the political arena, talk is cheap. If an environmental group successfully curtails oil exploration, it captures the value associated with the pristine environment but bears none (or only a small share) of the opportunity cost of forgone energy production. An oil company that is allowed to explore for oil with few regulations on environmental impacts faces lower costs of development because these costs are borne by those who value the pristine environment. Neither side faces the opportunity cost of their competing uses, and both are locked into a zero-sum game in which one party's gain is the other's loss. In a political setting, tradeoffs between the environment and energy are necessarily confrontational.

Recognizing that politics will determine the distribution of values from the energy and environmental resources, both sides will invest time and money lobbying government.[2] But lobbying does not create wealth; it only redistributes it. With many energy resources in the hands of state and federal bureaucrats, the transfer game has become extremely important for both oil companies and environmental groups, which realize that their wealth will be determined by bureaucratic decisions.

On the supply side of these wealth transfers are political entrepreneurs. Like their counterparts in the marketplace, political entrepreneurs seek opportunities to provide benefits for their constituents, but unlike their counterparts, political entrepreneurs do not necessarily face all of the opportunity costs of their choices.

Consider the difference between a private and a political decision about the location of a drilling platform. In the former case, suppose that a 100,000-acre ranch with the mineral rights is owned by a family with a strong environmental ethic. If the family is approached by an oil company that believes there is great potential for oil production on the ranch, the family will have to consider the potential profits from energy development and the environmental impacts. As long as the rights to explore and produce oil on the land are well defined and enforced, it is unlikely that the parties will battle over the right to drill, even though strong opposing preferences may prevail. The owner of the ranch has veto power over energy development, but a decision not to allow drilling carries an opportunity cost in the form of forgone royalties. The energy company has an incentive to keep the ranch owner happy by offering to develop the site in ways that will balance potential impacts on energy production with environmental values; otherwise, the company is unlikely to get permission to explore.

In contrast, if the decision to drill is on public land, there are no clear owners of the resources. Environmental interests get no royalties if development is allowed, so the cost of stopping development is zero. And energy companies have less incentive to be accommodating on environmental issues because it increases their costs without necessarily increasing the likelihood that they will be able to develop the land. In political decisions, one side's losses are the other side's gains, in contrast to private decisions, in which there are gains from trade. Hence in the private decisions, cooperation replaces political conflict as both sides prospect for harmony.

THE POLITICAL GAME

With political choice ruling our federal lands, millions of acres have been closed to prospecting for harmony. The amount of federal onshore acres that have been permanently removed from energy exploration by wilderness legislation has grown from 9 million in 1964 (the year the Wilderness Act was enacted) to over 104.5 million in March 2000.[3] This is not the whole story, however. Lands recommended for wilderness are off-limits as well. As of December 1994, the Congressional Research Service reported to Congress that there were over 29 million acres that were recommended for wilderness designation (see Table 7.1). When we include other lands effectively off-limits to energy exploration and development, such as wilderness study areas, most national parks and wildlife refuges, Alaska set-asides, and miscellaneous federal lands encumbered for conservation purposes by legislative and administrative restrictions, the total acreage swells to just over 271 million, or nearly 44 percent of the total onshore federal lands managed by the Forest Service, the National Park Service, the BLM, and the Fish and Wildlife Service.[4]

Much the same pattern has occurred on offshore areas. Not long after the Reagan administration unveiled its five-year leasing program in 1982, a series of congressional moratoriums removed more than 460 million federal offshore

Table 7.1 Wilderness Acres Managed by the Federal Government

	Designated Wilderness	Recommended for Wilderness
BLM	5,205,972	8,500,552
F&WS	20,685,372	2,020,573
FS	34,714,303	4,239,098
NPS	43,148,948	14,426,714
Total	**103,753,950**	**29,186,937**

Source: Ross W. Gorte, "Wilderness: Overview and Statistics," Congressional Research Service Report for Congress, December 2, 1994, on-line version, http://www.cnie.org/nle/nrgen_5.html.

acres from the program, including nearly all of the best prospects for major new offshore discoveries outside the central and western Gulf of Mexico.[5] On June 12, 1998, President Clinton signed an order extending the moratoriums "for more than 10 years and permanently protecting national marine sanctuaries from oil and gas drilling."[6]

The opportunity costs of removing areas from energy development may be quite high. For example, a 1986 Department of the Interior study reported an expected present value of the net economic benefits from petroleum development in the Arctic National Wildlife Refuge of $2.98 billion.[7] Alternatively, using updated oil price projections and revised tax and financial assumptions, a simulation model[8] applied by the Wilderness Society projects an expected present value of net economic benefits between $0.32 million and $1.39 billion.[9] The lower values in the Wilderness Society study result from an assumption that there will be a larger number of dry holes. Given the enormous differences in projected net benefits, reaching agreement on energy potential will be extremely difficult.[10]

The energy debate is made more acrimonious by the fact that decisions of whether to allow exploration most often involve hundreds of thousands of acres, with no provision for developing small increments of land under carefully controlled conditions. In this setting, it is not surprising that cooperation between environmentalists and developers is virtually nonexistent.

As a result, federal oil and gas leasing has fallen victim to drawn-out legal disputes that add to energy costs, as federal environmental assessments of proposed exploration and development are challenged in the courts.[11] Under the 1969 National Environmental Protection Act, federal agencies are required to prepare an environmental impact statement (EIS) for any "major federal action significantly affecting the quality of the human environment."[12] The EIS requirement alone resulted in lawsuits, lengthy delays, and judicial set-asides.

Even though development brings increased economic activity, higher incomes, and a higher tax base, local constituencies still oppose development. The reason for opposition is that benefits must be weighed against the costs of public services, including increased police and fire protection, roads, and hospitals. For example, in their 1981 study of the regional impact on Alaska of federal offshore oil and gas leasing, Porter and Huskey estimated that the total net burden on the people living in the area before leasing began was between $917 and $2,309 per capita.[13] And this burden often increases when oil development activities decline. Given such a tradeoff and given the risks that drilling poses to commercial fishing and tourism, local opposition is understandable.

Federal policy has been extremely ineffective in mollifying state and local opposition to offshore activity. Since 1983, the Georges Bank area has been closed to leasing by congressional moratorium. Leasing sales off the California coast also have been held up since 1984 by congressional moratoriums and litigation. And offshore exploration was ground to a halt in Alaska by lawsuits brought by coastal communities and environmentalists.[14]

Since the federal offshore leasing program began in 1954, nearly 38 million offshore acres—only 4 percent of the federal offshore acreage—have been leased.[15] During the early years, this small fraction may have been due to low energy prices and technological constraints. Since the early 1970s, however, significant improvements in technology have made ocean drilling more profitable. Nevertheless, the program continues at a snail-like pace compared with foreign offshore efforts. Over a 20-year period beginning in 1964, the United Kingdom leased 66 million acres of offshore area (including the huge oil discovery in the North Sea), and Canada leased some 900 million offshore acres.

Current policy under the 1972 Coastal Zone Management Act may do more to exacerbate problems with offshore leasing. The act is, in part, a grant-in-aid program, making available federal funds and technical assistance for the development and implementation of state coastal zone management programs. During fiscal years 1974 through 1982, the federal government distributed more than $489 million to states and territories to help them develop and implement their coastal zone management programs.[16] The legislation also offers states leverage over federal leasing decisions. Section 307 of the act and its implementing regulations require oil companies seeking a federal license or permit to carry out activities that affect land and water uses in a state's coastal zone to present an approved plan before it can obtain a "federal consistency certification."[17] This document must state that the proposed activity complies with and will be conducted in a manner consistent with the program of the affected state. The same requirement applies to a company seeking Department of the Interior approval of an offshore exploration or production plan.

The consistency provision has been used to delay offshore lease sales. In a 1982 lawsuit, the state of California contended that even the act of issuing a lease sets into motion a chain of events that results in oil and gas development, which directly affects the state-controlled coastal zone. A federal district court and the court of appeals for the Ninth Circuit agreed and ordered the Department of the Interior not to issue certain disputed leases in the Santa Maria Basin on which bids had been opened in May 1981. On January 11, 1984, the U.S. Supreme Court reversed the lower courts by ruling that the mere act of issuing a lease has no direct effect on coastal areas and that a consistency review is not required before a lease can be sold.[18]

Opposition to offshore leasing stems primarily from the costs and risks imposed on coastal communities. Development does entail local benefits of increased economic activity, additional jobs, and a higher tax base, but this may not be enough to offset the higher costs of public services that accompany such activity, as Porter and Huskey found for Alaska's offshore development program.[19] Victim compensation from public funds and the protection of operators through bankruptcy produce incentives that further lessen vigilance over safety. Until third parties have some assurance that strict liability will be enforced, opposition to offshore drilling and oil shipping is likely to persist.

Sharing oil and gas revenues among the federal government, coastal states, and local communities has the potential to mitigate state and local opposition to offshore drilling. Communities must be compensated in some way to offset the increases in infrastructure, police and fire protection, schools, and so forth that accompany a rise in offshore development. There are proposals to share offshore development revenues with states, but this revenue sharing must be extended to local communities that bear the costs of additional public services from offshore activities.[20]

THE POTENTIAL FOR SENSITIVE DEVELOPMENT

Ironically, at the same time that restrictions on and delays of federal leasing are increasing, there is growing evidence for the viability of environmentally sensitive development of oil-producing areas. For example, James Knight, Jr., has found that exploration and development caused "only short-term localized impacts" to elk in Michigan's Pigeon River Country State Forest. The elk tended to return to drilling areas two weeks to a month after drilling was completed, and they tolerated production more than they did exploratory drilling. Lightly traveled roads did not appear to threaten the animals.[21] Knight also found that seismic investigations caused more disturbance to elk than exploratory drilling and that serious impact was likely if the animals were disturbed during mating and calving periods, suggesting that timing can also mitigate harm.

Another set of studies focused on caribou and moose. Research on caribou began several years ago out of concern that migration routes and calving grounds would be irreparably harmed by the Trans-Alaska Pipeline and by energy development in Prudhoe Bay. Although these studies have not resolved the controversy on Alaska's North Slope, there are indications that, although oil and gas operations have altered animal behavior and movement, they have not significantly reduced caribou populations, as predevelopment critics had predicted. In fact, caribou populations have increased, and the moose have not been harmed.[22] Studies of other Arctic wildlife, including bears, small mammals, birds, and fish, have shown some habitat shifts, but no significant population changes. Certainly the level and approach of these operations must be carefully monitored and controlled to ensure the safety of wildlife in the Arctic region, but it is important to recognize that a low-disturbance operation is possible.

Numerous procedures can be used to maintain the environmental integrity of drilling sites and the surrounding area. Extremely fragile areas, such as marshes, can be protected through directional drilling, a technique that allows drilling on a site at some distance from the target. The Petite Anse 82 well on Avery Island, which is privately owned by the McIlhenny family, was drilled vertically for about a mile, then angled horizontally for about 12,000 feet to protect the surface environment.[23] At the drilling site, the fluid used to lubricate the drill bit and circulate rock cuttings to the surface was used to seal off and protect the penetrated rock strata from contamination and to control pressures in the well. Protective pipe and casings were used to prevent contamination of aquifers and other strata within several hundred feet of the surface. The space between the casing and the rock sides of the hole was filled with cement to prevent communication between strata, ensuring that freshwater zones were not damaged by zones containing saline water.

Other techniques are available to protect the surface area. Drilling-pad size can be minimized (from one to five acres) and self-contained. Noise and air emissions can be strictly controlled, safety valves and seals can be installed to stop the flow of unwanted fluids or gases from reaching the surface, and dikes and protective liners can be placed to control spills. If the well is dry, then the hole can be sealed, all equipment removed, and the site can be reclaimed, including restoring the original contours of the land, replacing topsoil, and reseeding with native vegetation. When all this is done, it is very difficult to find an abandoned well. For example, southwest of Yellowstone National Park, Getty Oil Company leased a site in an area known as the Palisades, in Bridger-Teton National Forest. This occurred just prior to the area being recommended for wilderness study. An exploratory well was completed in 1979, but no commercial quantities of oil were discovered at this site. After drilling was completed, the company worked closely with Forest Service personnel to reclaim the site, including revegetating the access road that had been built in extremely rugged terrain with slopes exceeding 70 percent in places. In 1981, Shell Oil Company reexcavated the reclaimed road and drilled a well a quarter mile from the original Getty well. That well was also a dry hole. The area was again reclaimed, the original contours reestablished, topsoil replaced, and native vegetation reseeded. Today, there is virtually no indication that either well was there. According to one Forest Service engineer, the site is a "showcase" reclamation project, where "you have to know where it is to find it."[24]

In offshore development, tanker operations pose the largest risk of damage to the marine environment. Between 1974 and 1982, there was an average of 17 accidents per year, with cleanup and damage costs over $250,000 (in 1987 dollars); and between 1974 and 1988, 7 accidents cost over $10 million. Before 1989, one of the costliest spills was the Amoco Cadíz spill off the coast of Brittany, France, for which Amoco was ordered to pay $85.2 million in damages.[25] On the positive side, the long-term effects have not been as bad as expected. In ten years, the coastal environment has recuperated quite well on its own.

The costliest transport oil spill in the United States occurred in the spring of 1989 when the Exxon Valdez spilled 10,836,000 gallons of oil in Alaska's wildlife-rich Prince William Sound. Exxon spent $1.28 billion (after-tax cost) on cleanup efforts but recovered only 1,604,000 gallons of the oil.[26] In addition to cleanup costs and various private claims, a settlement among the state of Alaska, the United States, and Exxon was approved on October 9, 1991. Exxon was fined $150 million, the largest fine ever imposed for an environmental incident. The court forgave $125 million of that fine in recognition of Exxon's cooperation in cleaning up the spill and paying certain private claims. Of the remaining $25 million, $12 million went to the North American Wetlands Conservation Fund and $13 million went to the National Victims of Crime Fund. As restitution for injuries caused to fish, wildlife, and lands, Exxon agreed to pay $100 million, which was divided equally between federal and state governments. In the civil settlement, Exxon agreed to pay $900 million, with annual payments stretched over ten years. The settlement included a provision allowing the state and federal governments to make a claim for an additional $100 million to restore resources damaged, the nature of which was not anticipated from the data available at the time of the settlement.[27]

In contrast to oil transport operations, the risk of a major oil spill from a platform or pipeline accident is extremely small. Offshore development using modern drilling and safety technology began in earnest in U.S. waters following the end of World War II and continued until Congress began declaring a series of moratoriums between 1982 and 1984. Between the end of the war and 1984, more than 30,000 wells were drilled in state and federal waters, with only one major platform accident resulting in significant amounts of oil reaching shore—the 1969 Santa Barbara oil spill.[28] The spill was traumatic and costly, causing damages of more than $16.4 million (1969 dollars), but again, the long-term environmental consequences have not been negative as expected.[29] Two more significant offshore accidents have occurred in U.S. waters since 1984. Both were the result of pipeline leaks and not the fault of the platforms themselves. Both spills were quickly cleaned up, and only moderate damage occurred.[30]

One reason for the good safety record of offshore platforms is the development of blowout preventers now used to close off wells if an unexpected change in well pressure occurs. Generally, a series of three or more blowout preventers is connected to the top of the well casing string. Another safety device consists of choke and kill lines, which permit drill operators to control pressures in a well if a blowout is threatened. These lines enable the operator to alter drilling fluid composition and flow in the well.[31]

In addition to high safety standards, expanding research and field testing indicate that offshore oil structures can offer something more than environmental coexistence. Fishery biologists know that fish can be attracted in marine waters by installing artificial reefs, and the construction and placement of such structures are common practices around the world. The Japanese, for example, have been building artificial reefs for over two hundred years to improve coastal fishing. Tenneco towed a retired oil platform from its original site south of Morgan City, Louisiana, to a site 275 miles southeast of Pensacola, Florida, and sank it in 175 feet of water. The Tenneco platform affords a high profile that attracts midwater as well as bottom-dwelling fish and a large surface area for attachment sites for crustaceans. In addition, active platforms in offshore Louisiana and southern California waters have proven to be excellent attractors for fish and fishers. Some of the most diverse marine communities in the Gulf of Mexico congregate around these platforms. Oysters, clams, lobsters, crabs, snappers, and many other species desired by commercial and sport fishers can be found in abundance at the base of these platforms.[32]

Given the good safety record of offshore platforms and the potential fishery benefits from them, restrictions on offshore drilling may be misplaced. Indeed, to the extent that these restrictions prevent domestic development and encourage greater oil imports, more petroleum will be transported by tankers, thus increasing the environmental risk, which can be much higher, from this source.[33]

PROMOTING HARMONY

Given the zero-sum nature of the tradeoffs in energy development on public lands, what guidance does free market environmentalism have for encouraging cooperation? The solution requires the specification of tradeable property rights that will force both sides to consider the opportunity costs of their actions. The best example of how this can work comes from the National Audubon Society's Paul J. Rainey Wildlife Sanctuary, a 26,000-acre preserve at the edge of the Intracoastal Waterway and Vermillion Bay in Louisiana. The Audubon Society, which opposed oil and gas development in ANWR and other public wilderness settings, acted differently when it owned the land and mineral rights. Because the Rainey Wildlife Sanctuary is privately owned by the Audubon Society and home to deer, armadillo, muskrat, otter, mink, thousands of geese, and many other birds, you would think it unlikely that Audubon would allow energy development there. But since the early 1950s, the sanctuary has been producing natural gas from 37 wells (and a small amount of oil), even though they are a potential source of pollution. These wells have produced more than $25 million in revenues for Audubon, but these revenues are not as high as they might be because Audubon also cares about the environment and imposes costly contractual restrictions on how the gas can be extracted.[34]

Rainey is not the only example of cooperation with industry in the Audubon system of private sanctuaries. Oil exploration and discovery brought the Michigan Audubon Society's Bernard W. Baker Sanctuary to the attention of Mobil Oil Corporation. In 1975, Mobil Oil approached Michigan Audubon with a proposition to explore and drill for oil in the Big Marsh. They offered a potential income of up to $100,000 per well with the possibility of four or five wells. Fiscal pragmatists in Audubon argued that Mobil offered a solution to the society's growing deficits, while others argued that the nesting grounds for the sandhill crane could not be disturbed at any price. When the members voted in October 1976 on a proposed amendment to the bylaws that would have prohibited mineral extraction in all Michigan Audubon Society sanctuaries, a majority of the 2,301 votes favored the amendment. But the majority was not the two-thirds necessary for passage. Then Mobil Oil withdrew its offer.

Five years later things had changed. An article in the May 1981 issue of Audubon describes the scene at Big Marsh Lake:

> They had just broken ground for the drilling pad, last time I went out to Big Marsh Lake. . . . Three or four weeks probably, and a sky full of sandhill cranes would be splashing down out there in the marsh. That's why the hardhats were in a hurry. They had to get the pad in, and find what they were looking for beneath the marsh, and get out themselves for a while, before the cranes returned. That's the way it was written in the contract. There was this timeclock, and when the cranes punched in, the hardhats would have to punch out.[35]

Given an operating deficit of $14,000 and the technical capabilities of directional drilling, the Michigan Audubon Society decided to allow Michigan Petroleum Exploration to explore the marsh. Michigan Petroleum was required to drill from a pad a half-mile from the marsh, use high-efficiency mufflers to minimize noise, contain drilling fluids, and finance studies of possible environmental problems. The society expected and received royalties of approximately $1 million, probably less than they could have obtained had they not demanded strict environmental controls.

In a conversation with Audubon's David Reed, manager of the refuge, John Mitchell captured how thinking has changed:

> We talked into the evening, Reed and I. He said he had come a far piece from Earth Day. I said, so had I. Once in an unguarded moment, he allowed as how he liked the idea of cooperating with industry in a situation where it was likely there would be no adverse impact on the biotic community. And I said that maybe if that kind of situation wasn't on the scarcer side of rare—well, then probably we would find more preservationists behaving like pragmatists. Or at least beginning to think that way.[36]

As we have noted, examples of "cooperating with industry" are hard to find on public lands where bargaining is between politicians rather that private owners. The problem for public resource managers is that it is difficult to know what values are being traded off in development decisions. Basically, the federal lands (on or offshore) produce two categories of goods: (1) commodities, such as timber, forage, oil, gas, coal, minerals, and commercial fish, and (2) amenities, such as wilderness, free-flowing rivers, endangered species habitat, marine sanctuaries, and other environmental values. Making rational tradeoffs using land for production of these two types of goods requires information on values. For the first category, obtaining values is relatively easy because the goods are sold in the marketplace. Getting that information for the second category, however, is more difficult because they are not marketed.

To resolve this tradeoff between energy production and environmental amenities on onshore lands, free market environmentalism offers options ranging from variations on leasing to complete privatization. Many researchers and commentators have detailed the advantages and disadvantages of these options, but most have focused on commodity leasing.[37] If market-based values for environmental amenities are to enter the equation, the leasing and sale options must be expanded to allow environmental groups to lease lands. Under current leasing arrangements, the only way environmentalists can get their voice heard is through politics. If environmental groups could compete to purchase or lease public resources, then their voices might be heard in the marketplace. For example, when a federal tract is opened for energy leasing, environmental groups should be allowed to bid. The groups may want to preclude all development on the land, or they may want to find ways of "cooperating with industry."

A common objection to this proposal is that environmental groups cannot compete with giant corporations that can profit from the resources. This argument misses two points. First, purchasing the equivalent of conservation easements on energy resources may not be that expensive. Groups can either accept lower royalties if they win the lease or they can pay development companies to use more costly methods that mitigate environmental impacts. This may be far less expensive for both sides than expensive litigation or lobbying. Second, environmental groups and their members are not necessarily poor. Table 7.2 shows the annual budgets for a number of large environmental groups. Allocating funds in excess of $1 billion to leasing or purchasing public lands could give these groups much more control and reduce the acrimony.

Another argument against market competition for public energy resources is that a value cannot be placed on environmental amenities. But when the government reserves land from development, a value is being placed on the reservation; the political process decides that the environmental values are worth at least the opportunity cost of the energy resources left undeveloped. Moreover, expenditures on litigation and lobbying reflect a minimum value for the environmental amenities in question. These expenditures are not insignificant. In 1997 and 1998, six environmental groups had over $2 million in lobbying expenditures alone (see Table 7.3). If these groups were using their funds for market competition instead of for transfer activity, then market prices would reflect these values.

A third argument against placing public resources in private hands is that there is a free-rider problem associated with many amenity values. For example, if people who never visit a wilderness area derive pleasure from knowing that it exists, then it would be difficult to directly charge them for the existence value they enjoy. The logic of the argument may be correct, but the sizes of contributions shown in Table 7.2 suggest that people are not always looking for an environmental free ride. These contributions may understate the value of the resource, but there is no way of knowing whether the efficiency effects of an understatement will be greater than those of an overstatement in the political arena, where the provision of amenities is subsidized.

Table 7.2 Annual Revenue for Selected Environmental Organizations in the United States

Organization	Annual Revenue	FY
The Nature Conservancy	$563,566,000	1998
Ducks Unlimited	103,547,351	1998
National Wildlife Federation	97,585,000	1998
National Audubon Society	72,206,456	1998
Sierra Club and subsidiaries	48,511,500	1997
Natural Resources Defense Council	27,425,429	1997
Environmental Defense Fund	25,805,496	1997
Greenpeace, USA	18,473,000	1997
National Parks and Conservation Association	17,687,857	1998
The Wilderness Society	14,663,837	1998
Defenders of Wildlife	11,984,550	1997
Trout Unlimited	5,460,000	1997
Izaak Walton League	2,986,275	1997
Total	**$ 1,009,902,751**	

Figures taken from annual reports.

Table 7.3 Lobbyists Spending* in Washington, D.C.

	1997	1998
National Wildlife Federation	$550,000	$220,000
Sierra Club	165,000	130,000
Natural Resources Defense Council	289,000	280,000
Defenders of Wildlife	40,000	140,000
Ducks Unlimited	52,000	40,000
Trout Unlimited	40,000	80,000
Total	**$ 1,136,000**	**$ 890,000**

* Lobbying data were compiled using disclosure reports, amendments filed, and/or the Lobbying Disclosure Act of 1995.

Source: Center for Responsive Politics, "Lobbyists Spending in Washington," March 10, 2000, website: http://www.opensecrets.org/lobbyists/98lookup.htm.

Although leasing or selling public lands is the preferable free-market alternative for harmonizing tradeoffs, there is an alternative—the establishment of a trust authority for managing the resource—that might overcome objections to a competitive auction pitting energy developers against environmental groups.[38] Consider the Grand Staircase-Escalante National Monument established by President Clinton in 1996. In its simplest form, a trust for managing the Grand Staircase-Escalante would assign certain powers to a board of trustees who would manage the monument's assets.[39] The trustees would have a fiduciary obligation to manage the assets within the constraints of the trust agreement. In this case the goal of the trust would follow that established by the Bureau of Land Management: "to protect a spectacular array of scientific, historic, biological, geological, paleontological, and archaeological objects." The trust would be required to cover all costs either from revenues generated from the assets in the monument or from private contributions of funds, property, or services by individuals, corporations, or charitable foundations. The trust would have a board appointed by the President of the United States, with staggered terms to overlap presidential elections, thus eliminating the possibility that any one President would immediately appoint a new board to do what he wanted it to. To ensure that the board of trustees would carefully balance multiple uses in the monument and consider the fiscal implications of its decisions, trustees would be nominated from interest groups. The interests represented would include environmental, recreational, wildlife, Indian, ranching, mining, oil and gas, and state and local government. Carefully structuring the mission and board representation would keep the trust committed to its objective and its beneficiaries.

Given the high value of the environmental, recreational, archeological, and paleontological resources and of the energy resources, the Grand Staircase-Escalante offers an excellent opportunity to use the trust approach to consider the tradeoffs. Table 7.4 shows the potential annual revenues from various activities in the monument. At $3,200,000, the value of energy resources is

Table 7.4 Potential Annual Revenues of the Grand Staircase-Escalante Monument

Recreation*	$4,302,000
Oil	1,250,000
Coal	1,950,000
Grazing	101,250
Total	**$ 7,603,250**

* Assuming $5.00 per visit.

Source: Grazing: Dennis Pope, Biological Team Leader, GSE, February 3, 1999; Recreation: Based on recreation visits provided by Barbara Sharrow, Visitor Services Team Leader, GSE, January 26, 1999; Commodities: Lee Allison, State Geologists, Utah Geological Survey, August 19, 1997.

significant. Under the current political structure, any efforts to develop the monument's energy potential will result in conflict, as has been the case with ANWR. If the board of trustees, however, had to consider the cost of saying no to oil, gas, and coal development and the potential for using revenues from such development to protect the amenity values of the monument, this conflict might disappear.

CONCLUSION

Oil and gas leasing as well as other commodity production on public lands has become a casualty of the zero-sum game. Environmentalists, development companies, and state and local interests have faced off against one another in the political process, where the stakes are high and winner often takes all. This arena fosters acrimony rather than cooperation among disparate users of natural resources.

Free market environmentalism, with its emphasis on well-defined and enforced property rights, provides an alternative. Where environmental groups such as the Audubon Society own energy resources in sensitive wildlife reserves, they are more willing to make tradeoffs because they face the opportunity cost of saying no to development. Allowing environmental groups to bid on potential energy leases is one way of forcing all sides to consider the opportunity costs of energy development, but many object to this on the grounds that environmental groups are not sufficiently well funded to enter this competition, even though the data on environmental group funding suggest otherwise. A more palatable alternative would be to consider a trust approach to public land management. Moving in this free-market-environmentalism direction would help get the incentives right by forcing trustees to consider the opportunity costs of saying no to energy development. In this way, we might avoid some of the acrimony that now permeates political land management.

CHAPTER 8

PRIMING THE INVISIBLE PUMP

Mark Twain supposedly quipped: "Whiskey is for drinkin' and water is for fightin'." In the arid West, where water is the lifeblood of agriculture, this adage has become especially appropriate, as municipal, industrial, and environmental demands for water have grown and generated competing uses. Traditionally, growing demands have been met by increasing supplies made possible with dams and canals, such as the Central Arizona Project, which cost billions of dollars to deliver water primarily to municipal users in Phoenix and Tucson.

For most of the twentieth century, the federal government subsidized the construction and maintenance of water storage and delivery projects designed to make the desert "bloom like a rose." The Bureau of Reclamation and the Army Corps of Engineers administer the use of water from these projects, providing nearly 90 percent of it to agricultural users, who pay only a fraction of what it costs to store and deliver. The artificially low prices for federal water promote waste of water supplies that are coming under increasing stress from industrial, municipal, and environmental demands. And despite these demands, the political allocation of federal water has been unresponsive, with few transfers of water made to other uses. We have learned from this political allocation system that if water runs uphill to money, it gushes uphill to politics.

But fiscal and environmental realities are forcing westerners to recognize that the days of solving water problems with concrete and steel are over. Former Colorado governor Richard D. Lamm described the change:

> When I was elected governor in 1974, the West had a well-established water system. . . . Bureau [of Reclamation] officials and local irrigation districts selected reservoir sites and determined water availability. With members of the western congressional delegation, they obtained project authorization and funding. Governors supported the proposals, appearing before congressional committees to request new projects, and we participated in dam completion ceremonies.

> In 1986, the picture is quite different. The boom in western resources development has fizzled. . . . Congress . . . has to worry about how to cut spending, not which [water] project to fund. . . . Farmers are trying to stay in business and are recognizing that their water is often worth more than their crops. Policymakers recognize that the natural environment must be protected because it is a major economic asset in the region.[1]

This political, social, and economic climate is bringing pressure to change the way water is managed. In the face of efforts to curtail government spending and protect the environment, the formal and informal institutions that govern water allocation must foster conservation and more efficient allocation of existing supplies and take water's growing recreational and environmental value into account. Water markets provide a way of doing this.

Free market environmental principles have become a coalescing theme among environmentalists and fiscal conservatives who oppose political water projects that are both uneconomical and environmentally destructive. This theme first manifested its potency in the defeat of the 1982 Peripheral Canal initiative, a project to divert northern California water to southern California. Opponents successfully convinced voters that high construction costs combined with high environmental costs associated with draining fresh water from the Sacramento delta made the project an economic as well as an environmental disaster. Following the initiative's defeat, Thomas Graff, general counsel for the California Environmental Defense Fund, asked: "Has all future water-project development been choked off by a new conservationist-conservative alliance. . . ?"[2] The answer appears to be yes, as construction of new, large-scale federal water projects has come to a virtual halt.

Water marketing can provide a basis for extending the alliance into the twenty-first century by encouraging efficient use, discouraging detrimental environmental effects, and reducing the drain on government budgets. Equally important, water marketing can release the creative power of individuals in the marketplace, enabling water users to deal with allocation problems specific to their demands and their local environmental constraints. As economist Rodney Smith explained, with water marketing "a farmer can apply his first-hand knowledge of his land, local hydrology, irrigation technology, and relative profitability of alternative crops to decide how much water to apply and which crops to grow on his land."[3]

THE PRIOR APPROPRIATION DOCTRINE

As with all aspects of free market environmentalism, water marketing depends on well-specified water rights; that is, rights must be clearly defined, enforceable, and transferable. Clearly defined and enforced water rights reduce uncertainty and assure that the benefits of water are captured. Transferable rights force users to face the full cost of water, including its value in other uses. If alternative uses are more valuable, then current users have the incentive to reallocate scarce water by selling or leasing it.

Unfortunately, well-specified and transferable water rights are often conspicuously absent from the legal institutions that govern water resources. Governmental restrictions produce uncertainty of ownership, stymie water transfers, and promote waste and inefficiency in water use. By removing these governmental restrictions and adhering more closely to the prior appropriation doctrine, water marketing can provide a mechanism for improving efficiency and environmental quality.

The prior appropriation doctrine evolved on the western frontier, where water was scarce and agricultural and mining operations required users to transport water considerable distances from the stream. Responding to the special conditions in the West, early California gold miners devised their own system for allocating water. They recorded each claimant's right to divert a specific quantity of water from a stream and assigned it a priority according to the principle of "first in time, first in right." Under this system, a market for water quickly evolved in the late nineteenth century, but it was short-lived.[4]

In 1902, the system for managing water in the West changed dramatically. The Newlands Reclamation Act ushered in massive subsidies for the storage and delivery of water and sent signals, especially to the agricultural sector, that water was cheap. The act established funding to construct and operate projects that would deliver water to arid western lands. Initially, western irrigators were to repay the construction costs within ten years of project completion. However, interest-free repayment schemes, together with deferrals and extensions of the period of repayment—and a dramatic rise in interest rates during the 1970s and 1980s—raised the value of the subsidy to as high as 95 percent of the actual costs.[5] In addition, the irrigation subsidy is not limited to free interest on repayment of construction costs. Although most projects are run for irrigation, operation and maintenance costs are subsidized by other uses, primarily hydropower.[6] Four examples of the subsidies to irrigation in the western United States are shown in Table 8.1.

Because nearly all reclamation projects are carried out with subsidized loans to water users, they are extremely well suited to pork barrel politics. The actual cost of dams and canals is spread over all taxpayers, but the benefits—the wealth that comes when dry land becomes productive—are concentrated among special interest groups. With the actual cost diffused, the average taxpayer is not well informed about the projects. In contrast, irrigation interests are keenly aware and

Table 8.1 Examples of Subsidies to Irrigation

Irrigation District (per acre-foot)	Irrigation Charge (per acre-foot)	Actual Supply Cost
Columbia Basin East	$ 4.19	$41.16
Glenn-Colusa	$ 1.46	$17.85
Oroville-Tonasket	$11.47	$21.33
Westlands	$15.80	$67.56

Source: Terry L. Anderson, "Water Options for the Blue Planet," in *The True State of the Planet,* ed. Ronald Bailey (New York: The Free Press, 1995), 279.

politically active. Politicians deliver water to these special interests, and they pay a fraction of the cost.

In addition, as irrigation investments, these projects are seldom economically viable. As shown in Table 8.2, only one project, Oroville-Tonasket, generates on-farm water value in excess of the actual supply costs. Expressed in different terms, if water were priced high enough to irrigators to recover construction costs plus interest in all four projects, only one would generate enough yield from irrigated farming to pay for itself and then some. The rest do not yield enough value to cover supply costs.

Despite making little economic sense, good pork barrel water projects are not easy to get rid of. For example, President Carter tried unsuccessfully to stop funding a "hit list" of federal water projects in 1979. As recently as 1993, Congress authorized completion of the Central Utah Project, which includes a series of dams, aqueducts, tunnels, and canals designed to collect water from the Colorado River drainage in Utah and transport it to the Great Basin.[7] The project will deliver water to irrigators at a cost of roughly $400 per acre-foot. The additional crops make it worth only $30 per acre-foot to farmers, but they pay only $8 per acre-foot, or one-fiftieth of the cost of delivery. Making little economic sense and heavily subsidized, the project must go on. Again, water gushes uphill to politics.

As if the fiscal implications were not enough, subsidized water projects encourage inefficient use and excessive demand. Paying only a fraction of the cost, irrigators have no incentive to consider alternative technologies and crops that would save water. In addition, many irrigation systems use less than half of the water that flows to them. The rest runs off fields, carrying with it pesticides, herbicides, and soil nutrients; evaporates as it moves through open canals; or percolates into the ground through unlined ditches. Lands can become waterlogged as farmers apply copious amounts of water to their crops.

These conditions can lead to serious environmental problems, as evidenced by what happened at the Kesterson National Wildlife Refuge. For years, California's Central Valley Project provided water to Westlands Water District farmers who paid only a fraction of the actual cost of delivery. This encouraged them to irrigate even marginally productive lands. Wastewater from these lands drained into nearby Kesterson refuge via a drainage system built by the federal government.[8]

Table 8.2 Project Viability—Value vs. Cost

Irrigation District (per acre-foot)	On-Farm Value (per acre-foot)	Actual Supply Cost
Columbia Basin East	$20	$41.16
Glenn-Colusa	$6	$17.85
Oroville-Tonasket	$90	$21.37
Westlands	$27	$67.56

Source: Terry L. Anderson, "Water Options for the Blue Planet," in *The True State of the Planet,* ed. Ronald Bailey (New York: Free Press, 1995), 279.

In 1983, the U.S. Fish and Wildlife Service noticed grotesque deformities in the birds and fish living in Kesterson. The toxic culprit was selenium that had leached from the soil and carried out to the refuge in the irrigation drainage from Westlands Water District. In small doses, selenium is necessary for life, but it can be a deadly pollutant when concentrated, as it was at Kesterson.

Stopping the flow of wastewater into Kesterson by shutting off water to the irrigation district might have been a solution, but Westlands' farmers and the banks who held the debt on their farms would have none of that. Instead, taxpayer-funded pollution control costing millions of dollars was implemented to solve an environmental travesty caused by water subsidies.[9]

Some of the problems could be eliminated if water were tradeable, but reclamation laws have generally restricted transfers because they would allow farmers to make windfall profits on their subsidized water. As a result, there are many questions about whether reclamation-project water can be sold or leased for a profit, whether it can be transferred away from the lands to which it was originally assigned, and whether it can be used for nonirrigation purposes. In 1988, the bureau made an effort to remove some of these barriers to transfers by declaring itself a "water market facilitator" and by outlining procedures to govern transfers of federally supplied water. These types of reforms suggest that water policy is moving in a direction that will get the incentives right by allowing markets to play a greater role in water allocation.

Since the 1980s, when economists and policy analysts began to recognize that water markets could help allocate water, we have come a long way. Trades between cities and agriculture have become more commonplace. Witness the trade between the Metropolitan Water District (MWD) and the Imperial Irrigation District (IID) in southern California. MWD constantly faces future water shortages because of the arid climate and growing population. For years, MWD could not acquire water from nearby agricultural sources because of restrictions on water transfers under California and federal reclamation law. A change in the California water code coupled with Bureau of Reclamation policy becoming favorable to water transfers in the region led to the 1989 Water Conservation Agreement. Under the agreement, MWD pays for ditch lining and other conservation projects within IID in exchange for the 106,100 acre-feet of water per year that is salvaged. The agreement is in effect for 35 years.[10]

MWD has carried out other water transfers that illustrate the variety of approaches water transfers can take. In 1992, MWD began a two-year fallowing program with Palo Verde Irrigation District farmers. In addition to getting $620 per acre per year from MWD not to grow crops, farmers get $315 per acre-foot from MWD for the 93,000 acre-feet of water it receives from the fallowing program. MWD also negotiated a dry-year option with the Arvin-Edison Water Storage District in California's Central Valley, under which MWD paid to store 115,000 acre-feet of its State Water Project entitlement under Arvin-Edison lands. In dry years, MWD also will pay Arvin-Edison to pump and use that water within the storage district in exchange for up to 128,000 acre-feet of the district's entitlement from the federal Central Valley Project. The cost to MWD of the water is $90 to $100 per acre-foot. MWD has carried out similar dry-year option agreements with the Coachella and Desert Water Agency.[11]

The story is similar in other parts of the West. For example, Utah's building boom and the requirement by cities that builders own adequate water rights to accommodate additional growth has led to an active market for water rights. Shares of water rights held by ditch companies were selling for as much as $3,200 per unit, a fourfold increase since early 1993. Along Colorado's front range, towns and cities have been buying water rights from ditch companies and irrigation districts and transferring the water to municipal lines. In Nevada, municipalities have been buying water rights from irrigators to secure future development. Las Vegas has a standing offer to buy water rights for $1,000 per acre-foot.[12]

Even at the Bureau of Reclamation, home of subsidized water use, the focus now is on transferring full ownership of federal water projects to local water users. Not only do current users through their local water districts assume complete responsibility for project operation, they also stand to gain directly from water trades with cities and other potential water customers willing to pay them a premium for any surplus water they generate. This cannot help but improve prospects for expanding water markets down the road. As of April 1999, Congress has passed two pieces of legislation designating title transfers of project water. The first (S. 538, P.L. 105-351) was enacted to convey the South Side Pumping Division of the Minidoka Project, located in Burley, Idaho, to the Burley Irrigation District. The second (H.R. 3687, P.L. 105-316) was enacted to convey the Canadian River Project to the Canadian River Authority in Texas, pending prepayment of the district's debt on the project. Other good prospects for transfer in the near term include the Clear Creek Unit of the Central Valley Project in California and the Pine River Project in Colorado.[13]

The market revolution in water is not confined to the United States. Chile, known for its application of market solutions to a variety of social problems, implemented a market-oriented water policy in 1974. The constitution of Chile, passed in 1980 and modified in 1988, reversed the expropriation of water rights by the state begun in 1966 and established secure, transferable water rights. The constitution states, "The rights to private individuals, or enterprises, over water, recognized or established by law, grant their holders the property over them."[14] Chile's market-oriented water policy is credited with giving farmers greater flexibility in selecting crops to respond to market demand, producing greater efficiency in urban water and sewage services, and allowing growing cities to buy water from farmers without having to buy land or expropriate water.[15]

Australian states have carried their water reforms a step further with interbasin water transfers. In 1992, the first interbasin water transfer involved a five-year lease of 7,982 acre-feet from a property on the Murrumbidgee River in New South Wales to a cotton farm on the lower Darling River in South Australia. Since then, the interbasin water market has developed incrementally through a series of temporary water transfers.[16]

The steady development of transborder water trading in Australia is the type of approach that is needed in North America. Unfortunately, proposed water transfers between the United States and Canada entail massive projects to deliver water from remote regions of Canada to populated areas of the United States. For example, the infamous North American Water and Power Alliance would have delivered 250 million acre-feet of water from northern Canada to the southern United States. The construction cost alone was estimated as high as $380 million in 1990 dollars. Projects like this encounter strong resistance from Canadian citizens who gain little or nothing as individuals. If such resistance is to change, trades will have to be scaled down and carried out incrementally, as was the case in Australia. [17]

While the push for grandiose schemes stymies international water trading, the prospects for water trades between states are looking up. Referring to prospects for water trades between Arizona, California, and Nevada, Interior Secretary Bruce Babbitt noted that "we are on the threshold of a new period in which the three lower division states together will regularly be using that full apportionment." He believes this will necessitate "encouraging conservation in use, voluntary transfers of water, and managing our storage and delivery systems so that we can meet the growing demands that the new century will bring."[18] At least with respect to individual states, the future of transborder water trading may be here.

EXPANDING WATER MARKETS

The same insights that have helped to refocus the debate on the efficacy of markets in the allocation of surface water can be extended to the more complex tasks of allocating instream flows and groundwater. America's environmental awakening and burgeoning demand for recreational facilities, coupled with the declining quality of many streams, have drawn attention to the importance of instream flows. In many areas of the West, the depletion of groundwater supplies and water contamination due to toxic wastes have raised new concerns about groundwater. Both instream flows and groundwater are thus prime candidates for free market environmentalism.

Instream Flows

At one time, the management of instream flows was restricted to the maintenance of flow levels sufficient for navigation and power generation; today, however, water must meet a broader range of instream uses. For example, adequate instream flow levels must be maintained to sustain fish and wildlife habitats. Maintaining adequate flow levels can assimilate pollutants that remain a threat to many inland water bodies. And there is a growing demand for instream recreational opportunities such as fishing and floating. With the value of instream flows rising, the problem is to facilitate reallocation from off-stream to instream uses, but most states have laws that significantly limit the option for this reallocation.

A major stumbling block to the private provision of instream flows are state laws that link diversion to beneficial use. In frontier mining camps, an appropriation could be made by anyone who was willing to use the water by diverting it from the stream. Claiming water for instream flows is further complicated by the "use it or lose it" principle.[19] According to this principle, if a water user is not diverting his water, it is available to others for appropriation. Therefore, if an environmental group were willing to pay a farmer to reduce irrigation diversions, the conserved water left instream would be subject to diversion by other irrigators.

Courts have generally made it clear that private appropriations for instream flows are not valid. In a 1917 ruling against a claim to appropriate water for a duck habitat, the Utah Supreme Court concluded that it was

> utterly inconceivable that a valid appropriation of water can be made under the laws of this state, when the beneficial use of which, after the appropriation is made, will belong equally to every human being who seeks to enjoy it. . . . We are decidedly of the opinion that the beneficial use contemplated in making the appropriation must be one that inures to the exclusive benefits of the appropriators and subject to his domain and control.[20]

In 1915, *Colorado River Water Conservation District v. Rocky Mountain Power Company* also emphasized diversion. The conservation district had sought to establish the right to appropriate instream flows for the propagation of fish, but in 1965 the Colorado Supreme Court found that there was "no support in the law of this state for the proposition that a minimum flow of water may be 'appropriated' in a natural stream for piscatorial purposes without diversion of any portion of the water 'appropriated' from the natural course of the stream."[21] As recently as 1979, instream flow claims in California were even denied to a state agency and a nonprofit public-interest corporation.[22] In both cases, the California Supreme Court argued that there was no evidence that there would be a diversion of or physical control over the water.

Having been denied the option of private appropriations, states have undertaken to reserve flows by other means. States have traditionally chosen to maintain instream flows by reserving water from appropriation, establishing minimum stream flows by bureaucratic fiat, conditioning new water permits, directing state agencies to acquire and hold instream flow rights, or using the public trust doctrine to establish that existing diversion rights do not trump the state's responsibility to maintain instream flows.

Regardless of which technique is chosen, protecting instream flows is difficult to carry out in the political arena. Traditional off-stream uses can dewater streams, which can "adversely affect and in some cases destroy valuable in-place commercial and recreational water uses."[23] Understandably, state agencies hesitate to reserve instream flows when such allocations would collide with existing diversion rights in fully appropriated (or nearly so) watershed. The water efficiency task force of the Western Governors' Association pointed out:

> States are reluctant to try to use their power to regulate to protect and enhance instream flow values on such streams because to do so may invite litigation. Additionally, where states have the authority to acquire existing water rights and to transfer them to instream flow rights, this authority has

> not usually been exercised because of budgetary constraints. . . . The gap in protection of instream flows on streams approaching full allocation and the absence of protection of these flows in some states, together with water code provisions that encourage consumptive uses, [leaves] instream flows only partially protected in western states.[24]

Given the problems with political water allocation, it is appropriate to ask whether water markets could do better.[25] Though traditional water rights have not allowed for instream flows, the rising demand for instream uses and new technologies for monitoring water use and water quality are putting evolutionary pressure on water rights. As law professor James Huffman explains, "Sophisticated technologies of stream flow monitoring can serve the law of instream flow rights just as the technology of barbed wire served the nineteenth-century law of private rights in grazing land. Defining the parameters of a right to instream flows is no more difficult than defining the parameters of a right to divert water for agriculture or industry.[26]

One important consideration in the definition of instream flow rights must be their impact on third parties. Particularly important is the effect that instream flow rights could have on off-stream water markets. To appreciate the potential impact of instream flow rights on transfers, consider the hypothetical Gallison River illustrated in Figure 8.1. At the head of the relevant portion of the stream, the flow is 2,000 cubic feet per second (cfs), and downstream are clearly specified water rights that fully appropriate the stream. For example, Farmer Shaw has a right to divert 500 cfs with a priority date of 1862, but only consumes 250 cfs of that diversion; the difference between diversion and consumption is her return flow equal to 250 cfs, and it is available to downstream diverters. This leaves 1,750 cfs flowing in the Gallison below Farmer Shaw. Farmer Hill has a right to divert 1,000 cfs with a return flow requirement of 450 cfs, and Farmer Stroup has a right to divert 500 cfs with no return flow. The town of Waterville has a claim to 500 cfs, with no return flow. The remaining 200 cfs crosses the state line and is claimed in the downstream state. Hence, the Gallison is fully appropriated; that is, the quantities of water claimed for diversion and consumption are well specified and just meet the claims of the water rights owners.

With rights so clearly specified, water marketing is possible. For example, suppose that Farmer Hill wishes to sell his water to Farmer Shaw, who wishes to move the right upstream to her farm. As long as Farmer Shaw diverts and consumes the same quantities as Farmer Hill, all existing water claims will be met. But suppose she plans to use Hill's water in a way that consumes all of his 1,000 cfs diversion. This reduces the flow downstream just above Stroup to 750 cfs and therefore leaves enough for Stroup, but it reduces the flow above Waterville and the state line to 250 cfs, an insufficient amount to meet the downstream rights. For this reason, water laws provide some mechanism, usually a contested case hearing, whereby potentially harmed water rights holders can protest transfers. In this case they would have no complaint if Shaw's diversion and consumption stayed the same as Hill's, but they would be harmed if she increases the consumption rate.

Figure 8.1
Hypothetical Gallison River

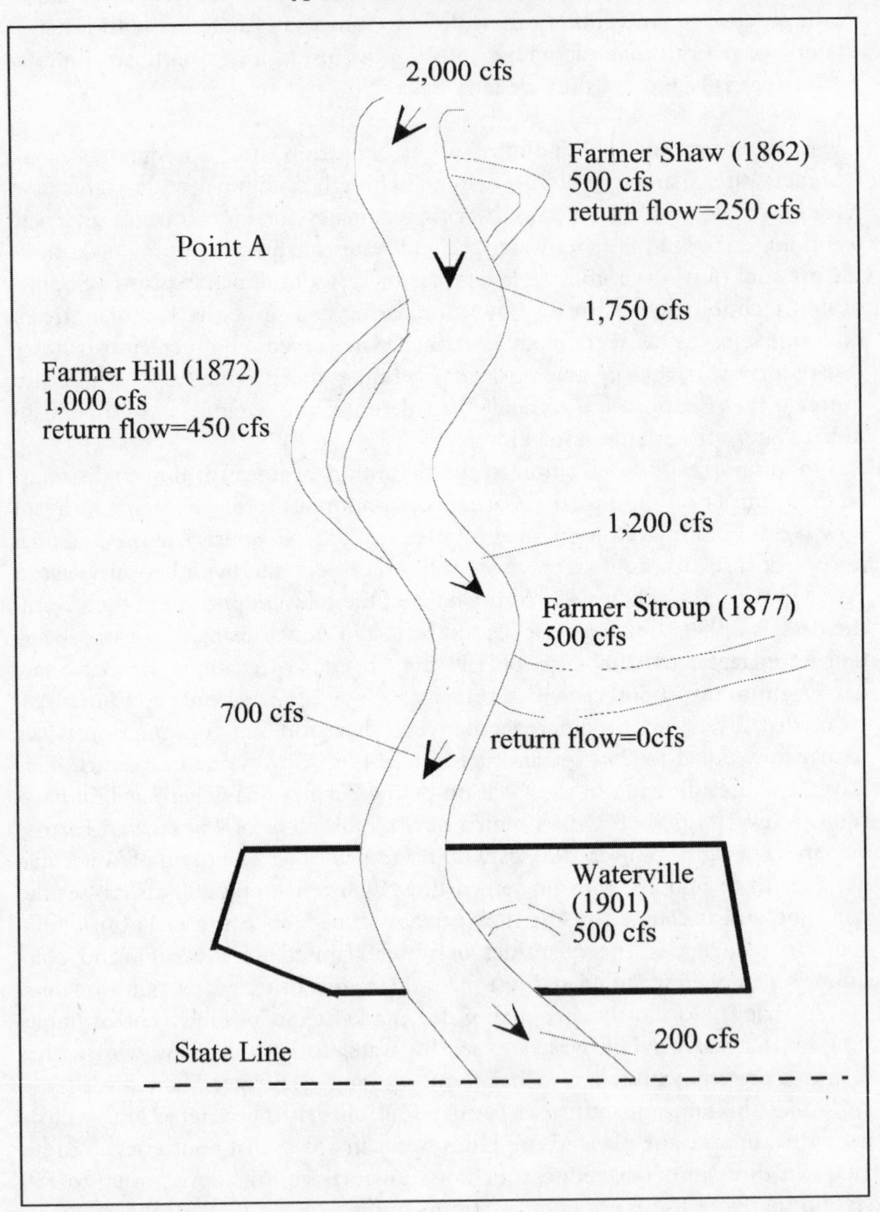

Now consider what happens to the potential transaction between Shaw and Hill if an instream flow claim to 200 cfs is established at Point A, between their two diversion points. In the absence of the instream flow claim, Hill can at least sell his consumptive right (550 cfs) to Shaw. But once an instream flow claim is established, any increase in consumption above that point will necessarily reduce instream flows and therefore impair the instream flow claim. Obviously, Hill has reason to oppose allowing instream flow claims because it limits his ability to sell or move his water right upstream, and Shaw has reason to oppose instream claims because it limits her options to buy water. The more instream flow claims there are and the longer the reach of the stream to which they apply (in other words, the claim could apply all the way to the state line instead of simply being at Point A), the more that instream flow claims limit the ability of existing water diverters to sell or move their water along the stream.

While instream flow claims can create constraints for existing diverters, parties wishing to establish instream flows through water markets can avoid these problems by purchasing upstream diversions and selling them to downstream users. For example, suppose that a group such as Trout Unlimited wanted to increase flows at Point A. It could purchase Farmer Shaw's water right and sell it to users in the downstream state. As long as other users cannot divert water between the point of purchase upstream and the point of sale downstream, the flows between the two points will be increased by 250 cfs. This actually happened during a 1984 drought in Texas, when resort owners along the Guadalupe River jointly purchased water from an upstream lake to increase the river's instream flow from 20 to 100 cfs.[27]

Environmental or recreational groups wanting more instream flows argue that allowing such market transactions still may not provide enough instream flows because it is too costly to prevent free riders from enjoying benefits such as scenic values, fishing experiences, and improved wildlife habitat. Nonpayers could reap the existence value, the critics point out; that is, they would have the satisfaction derived from simply knowing that an amenity exists, even if they do not consume or use it. For example, a Bostonian might be happy knowing that the Snake River in Idaho is free flowing and providing salmon habitat, even if she has no intention of looking at it, fishing in it, or rafting on it.

Of course, all goods have some potential for free riding, but the free-rider problem has not precluded the private provision of some instream flows. With increasing frequency, private groups and small communities have moved to secure protection and enhancement of instream flows. During the winter of 1989, for example, irrigators, the Nature Conservancy, the Trumpeter Swan Society, and the city of Grand Prairie, Alberta (Canada), contributed to such an effort on the Henry's Fork of the Snake River in Idaho. The resident population of trumpeter swans on the river was near starvation, its aquatic food supply cut off by river ice. Additional water from the upstream dam was desperately needed to clear a channel and allow the birds access to their food. A series of deals were struck between swan supporters and irrigators to increase flows to 400 cfs (up from 100 cfs), enough to break up the ice jam. To prevent recurrence, a long-term contract was also signed that guarantees a flow of 200 cfs, the amount necessary to prevent freeze-up.[28]

The market process can foster solutions to the free-rider problem because free riders represent opportunities for entrepreneurs who can devise ways of collecting from them. Environmental entrepreneurs in the Nature Conservancy, Trout Unlimited, and other private organizations play an important role in creating private rights and capturing the benefits of instream flows. With the right incentives, entrepreneurs in ranching and farming can accomplish similar results. Suppose that a rancher who owns riparian land is deciding whether to increase cattle grazing, which in turn would reduce fish habitat by destroying bank vegetation and causing siltation. Motivated by profit, the rancher is unlikely to give up grazing to preserve fish habitat unless he can profit from fishing.

For small streams, the cost of excluding nonpaying fishers is relatively low, making a fee fishing system viable. In the Yellowstone River valley, south of Livingston, Montana, spring creeks offer some of the world's greatest trout fishing, and sportsmen from all over the world try their hand at fly-fishing on those challenging creeks. Because the creeks begin and end on private property, many of the legal restrictions on private control of fishing access do not apply and upstream diverters do not exist. The private owners, who can collect a fee of between $50 and $75 a day from each angler, have a strong incentive to provide high-quality fishing. Grazing on the stream banks is limited and fishing access is controlled so that wildlife and land are protected, fish populations are sustained, and the fishers enjoy uncrowded conditions.

The use of markets has been more pervasive in England and Scotland, where scarcity has encouraged the development of property rights to fishing sites. With the demand for fishing opportunities rising, "there are few landowners . . . who can afford to ignore the commercial aspect of the sporting rights which they own."[29] It has become worthwhile for British landowners to incur the costs of specifying and enforcing contractual arrangements that govern fishing. As a result, many private, voluntary associations have been formed to purchase fishing access rights.

"In the 1960s and 1970s, smaller, privately managed fisheries that offered exclusivity in exchange for higher rod fees began to break out like an aquatic rash around [England]. Now every city and major town . . . has first-rate trout fishing within easy reach and at an affordable price."[30] In Scotland, "virtually every inch of every major river and most minor ones is privately owned or leased, and while trespassing isn't quite as serious a crime as first-degree murder or high treason, it isn't taken lightly."[31]

In Great Britain, angling is an accessory right to riparian ownership, and fishing rights can be leased to nonriparian landowners. The recent successes there suggest that the riparian doctrine still has merit as a means of limiting access to streams and protecting fish and wildlife habitats. The British system serves to enhance recreational opportunities, because such instream activities as fishing involve real opportunity costs that decision makers are forced to take into account when they consider the merits of water uses.

Instream-flow water markets are emerging as an important tool for enhancing instream flows in many parts of the western United States. Between 1990 and 1997, purchases, leases, and donations were reported in 9 of 11 western states, totaling more than 2.3 million acre-feet of water (Table 8.3). The federal government accounted for 70 percent of the water acquired, most of which was for enhancing flows on Idaho's Snake River to assist in salmon recovery. The remaining 30 percent was acquired by states and private groups.[32]

As of 1997, Oregon, Idaho, Washington, and California had more than 150 transactions totaling more than 1.7 million acre-feet of water acquired for instream purposes. Notably, these states implemented important legal or administrative changes to pave the way. In 1987, Oregon adopted changes to its water code that allows public or private entities to lease or purchase water rights and convert them to instream flows. As a result, the Oregon Water Trust, a private, nonprofit group, has made great strides in meeting its mission of securing adequate stream flows for spawning steelhead and salmon in the state. As of 1997, the trust increased flows on 25 different streams and rivers. In 1991, Washington established the Trust Water Rights program, which allows voluntary transfers of water for instream purposes. That same year, California changed its water code to allow water rights holders to change the use of their right for instream purposes. The following year Idaho granted exceptions to its water bank laws that allowed the Bureau of Reclamation to lease water from the water banking program for instream use. In 1995, an unlikely coalition of environmental and agricultural interests in Montana helped pass legislation that allows individuals or private organizations to lease water rights to ensure adequate flows for trout.[33]

Nevada and New Mexico have developed active instream flow markets as well. New Mexico is purchasing irrigation water rights on the Pecos River and retiring them to meet its flow obligation to Texas under a compact agreement. In Nevada, the Nature Conservancy spent $1.5 million to enhance flows to the Stillwater Refuge. The U.S. Fish and Wildlife Service has since taken over the task of acquiring water for the refuge. In 1997, the agency acquired over 8,000 acre-feet from local irrigators.[34] As of 1998, Great Basin Land and Water, a private water trust, has spent about $2.4 million to acquire 2,000 acre-feet of water to enhance flows on the Truckee River. The group got its start through the congressionally approved Truckee River Water Quality Settlement Agreement, which set aside $24 million for water acquisitions.[35]

Groundwater

As with past handling of instream flows, there has been practically no opportunity for the market to play a role in allocating groundwater in the United States. Free market environmentalism helps explain what has hindered market application and how markets could alleviate problems in the allocation of that resource.

Table 8.3 Western U.S. Water Acquisitions by Region, 1990–1997

Region	Quantity Acquired
Pacific Northwest	1,234,557 acre-feet
California	536,323 acre-feet
Southwest	306,658 acre-feet
Rockies	286,482 acre-feet
Total	**2,364,020 acre-feet**

Source: Clay J. Landry, *Saving Our Streams Through Water Markets: A Practical Guide* (Bozeman, MT: Political Economy Research Center, 1998), 8.

Like so many of our natural resources, concern is growing that groundwater sources are being depleted by increasing groundwater demands. In 1970, 19 percent of the water used in the United States came from groundwater sources. By 1985, groundwater accounted for 25 percent of the water used and provided drinking water for more than half the population.[36] Due to the rising demand, extraction now exceeds natural recharge in many key groundwater basins. One of the most dramatic cases of depletion is occurring in the Ogallala Aquifer, which underlies 174,000 square miles of the High Plains from South Dakota to Texas. Withdrawals from the Ogallala irrigate 15 million acres and account for 30 percent of total U.S. groundwater used for irrigation. Overdraft (withdrawals in excess of recharge) occurred in 95 percent of the Ogallala, sparking forecasts that the aquifer would be 23 percent depleted by 2020.[37]

In many parts of California, aquifers are overdrawn. Statewide, average annual groundwater withdrawals exceed average annual replenishment by 2 to 2.5 million acre-feet per year. The state has identified 11 critically overdrafted groundwater basins and 42 more basins where overdraft has taken place but is not considered severe.[38]

Land subsidence and saltwater intrusion are two problems that can result when water tables fall excessively. Pumping in the Floridan aquifer has drawn saltwater from the seaward edge of the aquifer along the coasts of South Carolina, Georgia, and Florida. In the area of Kingsville, Texas, pumping from numerous wells has led to saltwater intrusion in the Evangeline aquifer. Overdraft in the Houston-Galveston region has caused subsidence up to 8.5 feet in a 4,700-square-mile area. Intensive groundwater pumping in California's San Joaquin Valley has resulted in an area the size of Connecticut subsiding by as much as 30 feet in some places. In parts of Arizona, heavy pumping has resulted in land subsiding by as much as 12 feet and earth fissures that damaged property.[39]

Does such groundwater depletion make good economic sense? From an efficiency standpoint, the answer may be yes.[40] Mining a groundwater basin, as it is called when withdrawals exceed inflows, is appropriate if the future value of the water is expected to be lower than current value and there are no third-party impacts such as land subsidence.

A major difficulty in allocating groundwater, however, is that the future value of water left in a basin may not be available if there is open access to pumping. Water left for the future is often subject to the tragedy of the commons. Suppose that an individual must decide whether to leave water in a basin in order to offset future shortfalls in precipitation or surface water availability. Even if he believes that the current consumption value is less than the discounted future insurance value, his incentive to leave water in the basin is reduced by the knowledge that other users can pump the water immediately. Each individual realizes that anything he leaves behind may be consumed by others. And if others pump enough to lower the level of water in the basin, future pumping costs may be higher too. In the absence of secure ownership claims, future value gets zero weight in the calculus.

Groundwater users cannot optimize the rate of extraction unless the rights to water in a basin are clearly defined by water institutions and the courts. Only then can users accurately calculate the current and future value of groundwater supplies. Thus, the first step in solving the problem of depletion is to secure well-defined rights to groundwater use, which in turn will facilitate market transfers.

To gain an understanding of the present property rights structure, it is important to realize that the underpinnings of groundwater law are found in English common law. Because little was known about the hydrology of groundwater, rights to groundwater were assigned to the owners of the overlying land. As Frank Trelease has pointed out, "It was in the light of this scientific and judicial ignorance that the overlying landowner was given total dominion over his 'property,' that is, a free hand to do as he pleased with water found within his land, without accounting for damage."[41] When groundwater rights either are not assigned or are assigned on the basis of overlying land, the common pool problems can become severe. Each individual achieves the greatest net benefits by pumping water earlier than the others, because the lift costs increase as the level is lowered. The tragedy of the commons occurs because each individual has an incentive to pump water earlier than everyone else, so the supply is rapidly depleted.

Such poorly defined rights were harmless as long as there was little demand for groundwater; but as demand has increased, changes have occurred in the institutions that allocate groundwater, but these have not led to efficient use. For example, owners of overlying land have been granted rights to groundwater on the basis of reasonable use. The problem with the rule of reasonable use is that the interpretation of reasonableness has been subjected to the whims of judges and administrators, which has made the tenure of rights uncertain. In addition, the equal rights provision has usually been interpreted to mean that water use is restricted to overlying land and cannot be transferred elsewhere.

Conservation could be served if a system of property rights to groundwater were developed similar to the prior appropriation doctrine. Economist Vernon Smith has proposed how this could happen through issuance of property deeds for two types of groundwater rights: one is a share of the total water stock in a basin, and the other a share of the average annual recharge flow of the basin. The maximum initial allocations of each component would be proportional to users' pumping rates based on historical use. Smith developed his proposal in the context of the Tucson Basin in Arizona, using a base period of 1975. In this context, the initial allocations would be a function of the 224,600 acre-feet of water extracted during the year. If an individual had used $x_{(i)}$ acre-feet, his proportion, $P_{(i)}$, would be $x_{(i)}/224{,}600$. He would thus be entitled to receive a property deed for a stock right in proportion, $P_{(i)}$, to the Tucson Basin's total water stock, which was approximately 30 million acre-feet in 1975, and a property deed for a flow right in proportion, $P_{(i)}$, to the basin's average annual recharge flow, 74,600 acre-feet. Basing the initial allocations on prior use, however, could promote water waste, as potential rights holders would be encouraged to race to the pump house.[42] To prevent this, the initial allocations could be based on the proportion of a user's land that overlies an aquifer rather than on prior use.

In order to enforce the newly defined rights, metered pumps could be used and periodic readings taken. At the end of each year, an adjustment could be made by subtracting a user's share of the total recharge flow from the amount he had used. Stream flows or other sources of recharge could be used to estimate the total recharge flow for a given year. Those who pumped more water than they owned could be fined or could have their overdraw subtracted from future entitlements with interest charged.

Allowing transfers of property rights in stocks and flows would promote efficiency because users would have an incentive to compare the opportunity costs of various uses. Furthermore, groundwater users would have a greater opportunity to stabilize water levels if they could make exchanges with outside suppliers, which users of surface water in certain states are already permitted to do. They could adjust their stocks and flows by making purchases during dry years and by selling temporary shares during wet years. As with surface water exchanges, groundwater exchanges would have to be based on consumption criteria in order to protect buyers and sellers against third-party effects. An additional benefit of the exchange system is that the risk of dry years would be distributed among many producers and consumers. Risk-averse parties would have an incentive to acquire or hold greater shares in groundwater stock and flows. The water market would thus incorporate a voluntary savings plan that permitted users to guard against water shortages.

But the assignment of property rights in stocks and flows would not necessarily solve all the common-pool problems. A holder of a title to a stock of water could still face high extraction costs imposed by the usage rates of other pumpers. In addition, if the holder is located near the perimeter of a groundwater basin, the holder could find his well dry if the aquifer is drawn down by other pumpers. Such third-party effects could be minimized through a slight modification of Smith's property rights scheme. For example, when initial stocks are allocated to private owners, some stocks could be withheld from allocation to ensure that the basin's water level does not fall too low. Finding the ideal level of stocks to be withheld for the initial allocation, however, is difficult because of the uncertainty in estimating many of the hydrologic and economic variables. This will become easier as pumping occurs and generates better information on hydrological characteristics of the basin. Thus, barring significant economic or hydrological surprises, the economic costs of error in the choice of stocks withheld should be relatively small in the long run.[43]

Unitization, a contractual arrangement used to mitigate common-pool problems in oil extraction, is another way of achieving efficiency.[44] Under such an arrangement, all parties would contract to use agreed-upon methods of extraction and delivery and to share the costs. Some wells would be shut down and others would remain operational. Those wells left open would be strategically chosen to prevent the cone-of-depression problem, wherein pumping from one well draws water from adjacent wells. Each party's share of the lift costs would be based on his usage rate. For that reason, unitization might entail higher delivery costs, but it would also foster increased water conservation and thus lower lift costs.

Of course, because unitization would require the cooperation of all shareholders of groundwater, negotiating such agreements would entail higher transaction costs than negotiating two-party contracts. Increases in the value of groundwater and the cost of excessive water withdrawals, however, might justify proceeding with unitization, at least for small basins. What is important at this stage is to remove and to refrain from creating legal obstructions to the evolution of such arrangements. That means choosing a set of rules that would enhance the specificity, enforceability, and transferability of property rights.

As groundwater scarcity threatens pumpers with increased costs and limited supplies, well-defined property rights to groundwater basins become a necessity. Though limited to recharge rights, assignment of rights in the Tehachapi Basin in Kern County, California, has produced noticeable improvements.[45] For example, water levels in the basin are no longer falling, eliminating upward pressure on pumping costs. Incidental recharge has raised the water table in some areas. The city of Tehachapi no longer rations water as it did periodically before adjudication, and rising water tables have brought previously marginal wells into production. Finally, the fact that the rights adjudicated were transferable has facilitated reallocation to new uses. Agriculture was the dominant user of the basin before adjudication and held the lion's share of groundwater rights. Those rights are now leased and sold locally, mostly to cities and service districts that purchase both land and water rights for development. The resulting reallocation has been accomplished cooperatively through voluntary exchange.[46]

CONCLUSION

Dire predictions that the blue planet will face global water shortages may unfold if we cannot supplant water politics with water markets. Growing demands for consumption, pollution, dilution, and environmental amenities will put pressure on limited resources. But those pressures need not create a water crisis if individuals are allowed to respond through market processes. Perhaps more than with other natural resources, water allocation has been distorted by politics under the notion that water is different. Some would say that water cannot be entrusted to markets because it is a necessity of life. To the contrary, because it is a necessity of life, free market environmentalism argues that water is so precious that it must be entrusted to the discipline of markets that encourage conservation and innovation.

In order to reap the benefits of water markets, policy makers must stay the course. They must continue to find ways to define property rights in water, enforce them, and make them transferable and then guard against doctrines that erode these principles. The prior appropriation doctrine supports these principles, but the public trust doctrine erodes them. By limiting the application of the public trust doctrine, by extending the application of the prior appropriation doctrine to instream flows, by instituting clearly defined property rights to groundwater basins, and by reducing the impediments to exchange for nontraditional uses such as instream flows, policy makers will vastly improve water allocation.

CHAPTER 9

HOMESTEADING THE OCEANS

As one of the world's largest commons, oceans provide a challenge for free market environmental solutions. Outside the territorial limits of sovereign countries, only weak treaties limit the use of ocean resources for fishing, mineral or energy development, shipping, and garbage disposal. With few restrictions on entry, a tragedy of the commons can occur, resulting in such problems as pollution and severely depleted fish stocks.[1] The Food and Agriculture Organization of the United Nations reports that 25 percent of the commercial fish stocks in the world are overfished and another 44 percent are fully exploited.[2] Moreover, pressure on the commons is increasing, as new technologies raise returns to exploiting ocean resources. For example, new drilling techniques make deepwater oil exploration and production feasible; shipping technologies are increasing the size of oil tankers and the potential for oil spills; and far-ranging vessels equipped with sonar, onboard processing, efficient harvesting devices, and refrigeration allow fishing fleets to deplete ocean fisheries.

Ocean fisheries are also being exploited by recreational fishing. In 1996, 9.4 million anglers spent 103 million days catching cod, flounder, bluefish, salmon, striped bass, mackerel, and other popular game fish in U.S. marine waters. In fact, the catch of some species by saltwater angling enthusiasts has had a greater impact on population size than the activities of commercial fishers. Sport anglers, for example, are allocated 68 percent of the total allowable catch in the king mackerel Gulf of Mexico fishery. The growing influence of saltwater sportfishing, coupled with commercial harvesting, has intensified pressure on ocean fish stocks.[3]

These rising demands and new technologies are creating pressures to change the rules governing access to ocean resources. Historically, access to resources beyond narrow territorial waters (typically 3 to 12 nautical miles from shore) was open to anyone for the taking. Gradually, however, coastal nations have begun to exert greater control over resources lying farther off their shores. The move to fence ocean resources began in 1945, when President Harry Truman claimed that the United States had exclusive rights to mineral and hydrocarbon resources lying on or under its continental shelf. What followed was a steady procession of declarations by coastal nations to extend claims to resources lying within 200 miles off their shores. These claims have converted coastal waters from a regime of largely open access and high-seas freedom to one with significant national controls over resource uses.[4]

Like the evolution of property rights to land and water on the American frontier, extending territorial limits is not the final solution to the problem of open access to ocean resources. As economist Ross Eckert points out, the conversion of open access to limited access "does not guarantee the improved allocation of ocean resources" but is "only a first step for removing the inefficiencies that result from communal rights."[5] Over 20 years have elapsed since the United States extended its territorial limits to 200 miles from shore, but a significant number of fish stocks are still being overfished. In a 1998 report to Congress on the status of U.S. fisheries, the National Marine Fisheries Service (NMFS) categorized 90 stocks as being overfished and 10 stocks as approaching an overfished condition.[6] Table 9.1 lists some of these species.

In contrast to an earlier period, today's problem with overfishing cannot be blamed on foreign fishing. The foreign vessels that depleted fisheries in waters off the United States during the 1960s and 1970s have been mostly removed in an effort to encourage the development of the domestic fishing industry. Notably, domestic fishing effort did rise substantially during the 1980s and early

Table 9.1 Sample of U.S. Fish Stocks Being Overfished, September 1998

New England	Mid-Atlantic	South Atlantic
Atlantic sea scallop	scup	king mackerel
American lobster	summer flounder	(Gulf group)
Atlantic cod,	black sea bass	
Gulf of Maine	bluefish (except	Gulf of Mexico
American plaice	Gulf of Mexico)	red snapper
witch flounder		Nassau grouper
winter flounder,	Main Hawaiian Islands	jewfish
Gulf of Maine	pelagic armorhead	red drum
winter flounder,	squirrel fish snapper	
southern	longtail snapper	South Atlantic
New England		red drum
		jewfish
North Atlantic		Nassau grouper
swordfish		vermillion snapper
blue marlin		red porgy
white marlin		black sea bass
sailfish		

Source: National Marine Fisheries Service (NMFS), *Report to Congress: Status of Fisheries of the United States*, September 1998, website: http://www.nmfs.gov/sfa/98stat.pdf.

1990s, putting pressure once again on the reproductive capacity of fisheries in U.S. waters. After two decades of regulation imposed on our own fishers, the striking result has been the continued decline of key fish populations. As with reforms in land resource policy, the key to effective fishery policy lies in a property rights approach.[7]

THE OCEAN COMMONS

Ocean fisheries provide the classic case of the tragedy of the commons,[8] because many species of fish are mobile and access is difficult to monitor. Therefore, the rule of capture often dominates; any fish left by one fisher is available to another. Rather than leaving fish to grow and reproduce, the incentive is to harvest the stock before others do. With each fisher facing this incentive, the end result is for the fish stock to be overexploited. Whether the population of fish ends up becoming extinct ultimately depends on the cost of capturing the last fish in the stock. Because these costs tend to rise exponentially, declining fisheries have historically reached commercial extinction before biological extinction; that is, the additional costs of capturing the few remaining fish exceed the returns, so that it has become unprofitable to continue fishing.[9]

Nonetheless, open access to the resource frequently results in a lower than optimal (if not total depletion of) stock and an overinvestment in fishing effort. As long as the cost of taking an additional fish is less than the value of the fish, a profit can be earned. But with open access, not all costs will be taken into account. Another fish taken from the stock can reduce the reproductive capacity of the fishery and raise search and capture costs for other fishers. Because these added costs are external to an individual fisher who considers only his costs and benefits, over time there will be too many fishers in the fishery. In addition, open access encourages a rate of exploitation that will be too rapid. Being the first to exploit the fishery allows the highest returns, because the costs of finding and catching fish will be lowest. This race to the best fishing grounds is often manifest in the form of overcapitalization in radar, sonar, faster boats, and larger nets. The result is lower profits for the too many fishers investing in too much capital to catch too few fish.

Economist Frederick Bell provided one of the first empirical verifications of overexploitation of an open-access fishery. In his examination of New England's northern lobster fishery in 1966, he found that an efficient output of lobster would have occurred at 17.2 million pounds. To attain this output, the efficient number of lobster traps would have been 433,000. However, during 1966, Bell found that fishers employed too much capital—891,000 traps—to harvest too many lobsters—25 million pounds.[10]

For many of the world's ocean fisheries, government control has replaced no control. Today's tragedy is that government control has not prevented overexploitation in fisheries, but has greatly increased costs.

REGULATING THE OCEAN FISHERY

Instead of relying on a property rights solution, government regulation has been the traditional mechanism for controlling overexploitation of the ocean commons. Unfortunately, there are inherent problems with regulation because the regulators do not own the resources and do not face economic incentives to manage it efficiently. For the fishery, most regulatory schemes have focused on sustaining the maximum yield, that is, in allowing the largest quantity of fish that can be caught year after year without depleting the stock. Economists argue that this yield is usually not the yield that maximizes profits, however, because it ignores economic variables such as discounted returns of future catches and the costs of present and future extraction.[11]

On the whole, regulatory schemes focusing on maximum sustainable yield and ignoring economic factors have led to lower profits and economic wastes in United States fisheries. Regulatory policies in the United States before 1976, for example, attempted to reduce catch and maximize sustainable yield by raising the cost of fishing. In the Pacific salmon fishery early in the twentieth century, regulators prohibited the use of traps first perfected by the Indians, who caught the salmon when the fish returned to spawn. With the elimination of traps, fishers chased the salmon in the open oceans. The substitute for traps became very expensive, with sophisticated equipment that still allowed fishers to overexploit the resource. When the number of fishers and the length of the season were restricted, entrepreneurs bought bigger boats, sonar, and more efficient nets. To plug these holes in the dike, regulators then established other layers of regulations controlling seasonal limits. The salmon catch was ultimately curtailed, but the approach generated economic inefficiency, as more labor and capital were applied to catch fewer fish. As fishers were forced to fish longer in less productive areas with more expensive equipment, economic waste reduced the net value of the fishery. In 1976, Francis T. Christy, Jr., estimated that the overcapitalization and overuse of labor in American fisheries cost $300 million per year, or at a 6 percent interest rate, $5 billion in perpetuity.[12]

In addition to the overcapitalization, the regulatory process sometimes spurred absurd restrictions. For example, Maryland oystermen at one time could use dredges but had to tow them behind sailboats on all but two days of the week, when motorized boats were allowed. And in some Alaskan fisheries, fishing boats were limited to 50 feet in length.[13]

Such restrictions often favored one user group over another. After examining data on regulation policies during the 1960s and 1970s, economist James Crutchfield concluded that the regulatory process had "generated an ever-increasing mass of restrictive legislation, most of it clothed in the shining garments of conservation, but bearing the clear marks of pressure politics." The overwhelming majority of these restrictions, Crutchfield decided, reflect "power plays by one ethnic group of fishermen against another, by owners of one gear against another, or by fishermen of one state against another state."[14] The combined costs of regulations led Robert Higgs to conclude:

> The social resource waste has therefore grown steadily larger over time. Today, from a comprehensive point of view, the Washington salmon fishery almost certainly makes a negative contribution to net national product. The opportunity costs of the *socially unnecessary* resources employed there, plus the *socially unnecessary* costs of government research, management, and regulation, are greater than the *total value added* by all the labor and capital employed in the fishery.[15]

Meanwhile, despite these costs, many fisheries in U.S. coastal waters during the early 1970s were either in trouble or on the verge of it.[16]

The Magnuson Fishery Conservation and Management Act of 1976 tried to remove some of these regulatory inefficiencies by setting a new direction for fishery policy. It extended the nation's marine management jurisdiction from 12 to 200 miles offshore and encouraged the development of domestic fisheries. Eight regional councils were established with the authority to manage fisheries under their jurisdiction. Notably, the act does not mandate the standard of maximum sustainable yield, but rather stipulates that fishery management plans may "establish a system for limiting access to the fishery in order to achieve optimum yield."[17] Optimum yield in this case must take into account economic variables, such as interest rates, fish values, and the cost of alternative technologies.

While the Magnuson legislation was a step in the right direction, significant problems remain for regulators. The legislation encourages licensing entrants as a way of limiting the number of fishers or vessels in a fishery, but limiting entrants "cannot prevent crowding, congestion, strategic fishery behavior, racing, and capital stuffing."[18] Controlling the intensity of effort remains a thorny problem, because fishers are substituting fewer larger boats for more smaller boats. The result is that "rising fish prices constrained by a limited number of vessels, and unconstrained by any sort of territorial limit, has led to vastly increased individual fishing capacity."[19] Even with licensing, regulators find that a few powerful fishing vessels can do in a few minutes what used to take days. The bottom line is that the Magnuson Act has not rebuilt many declining fisheries in U.S. waters. According to a 1999 status report on U.S. marine resources by the National Marine Fisheries Service, 34 percent of the fish stocks of known status are overutilized—i.e., fished with excessive fishing effort—and many of these remain significantly below their historic levels and continue to be overfished (see Table 9.2).

Complicating the regulation of commercial fishing is recreational, or sport, fishing. With the impact of recreational fishing in U.S. coastal waters, several fish stocks have come under pressure, forcing regulators in some instances to take drastic action. For example, in 1987 the National Marine Fisheries Service banned commercial fishing of king mackerel after recreational fishers from Texas through southeastern Florida exceeded a 740,000-pound catch limit. Meanwhile, tension has grown between recreational and commercial fishing interests. Commercial fishers fear that recreational fishers under the guise of conservation have become highly influential in setting policy at both the state and federal levels. Recreational fishers complain that fisheries must not be managed solely for

Table 9.2 Representative List of Stocks, 1999

Geographic Area	Overfished	Overcapitalized
New England	Atlantic cod, silver hake, red hake, winter flounder, summer flounder, witch flounder, American plaice, goosefish, sea scallops, American lobster	Atlantic cod, silver hake, red hake, summer flounder, winter flounder, witch flounder, American plaice, goosefish, sea scallops, American lobster
Atlantic	black sea bass, scup, tilefish wolffish	black sea bass, scup, tilefish, wolffish
Gulf of Mexico	red snapper, Nassau grouper, shallow groupers	red snapper, Nassau grouper, shallow groupers
South Atlantic	vermillion snapper, red snapper	vermillion snapper, red snapper
Caribbean	Nassau grouper	Nassau grouper

Source: National Marine Fisheries Service (NMFS), *Our Living Oceans: Report on the Status of U.S. Living Marine Resources*, 1999, U.S. Department of Commerce, NOAA Technical Memorandum NMFS-F/SPO-41, website: http://spo.nwr.noaa.gov/olo99.htm.

commercial interests and that they have as much right to the resource as commercial interests do. The management districts for both the South Atlantic and Gulf Coast fisheries have experienced intense pressures from these two groups.[20]

The conflicts between commercial and sport interests have led to concerted efforts by both groups to gain stronger footholds in regional fisheries management councils. Historically, the regional councils of the National Marine Fisheries Service were dominated by representatives from commercial fishing interests. This dominance may be changing, as recreational interests have managed to gain more political influence. Meanwhile, the turf battles have left questions concerning the regional councils' effectiveness in managing fish stocks:

> As the battle between user groups has intensified, many observers have questioned whether NMFS and the councils, in their desire to satisfy every demand for a piece of the resource pie, have lost sight of their fundamental responsibility to protect the health of fish stocks.[21]

INDIVIDUAL TRANSFERABLE QUOTAS

A relatively recent innovation in fishery management that is an improvement over a strictly regulatory approach is a system of individual transferable quotas, or ITQs. An individual quota entitles the holder to catch a specific percentage of the total allowable catch, generally specified by a government agency. This system is attractive for several reasons. First, each quota holder faces greater certainty that his share of the total allowable catch will not be taken by someone else. Under the current system, an individual's share of the total allowable catch is determined by who is best at capturing the fugitive resources. With ITQs, holders do not compete for the shares, so there is less incentive to race other fishers. Second, transferability allows quotas to end up in the hands of the most efficient fishers—that is, those with the lowest costs and who can pay the highest price for the ITQs. Less efficient producers and inputs move to other industries. As a corollary, ITQs encourage progress in reducing the cost of catching fish and enhancing the quality of the fish delivered to markets. Fishers who adopt new cost-reducing or quality-enhancing methods make more money with their quotas and are in a better position to purchase quotas from those who are less efficient. This is in marked contrast to regulations, which encourage overinvestment in the race for fugitive resources.

New Zealand and Iceland employ ITQs as a major component of their fisheries management system, and both have experienced considerable success in reducing the race to catch fish. The deplorable state of many of New Zealand's inshore fisheries and a government favorable to market solutions led to the introduction of ITQs in 1986 into 29 commercial fisheries. Today, fish stocks are generally healthy, and the value of fisheries as represented by quota value has grown to nearly $1.4 billion.[22] ITQs are also encouraging cooperation among fishers in enhancing fish stocks. For example, overfishing before the introduction of ITQs decimated the paua (abalone) fishery. With ITQs, quota holders in the Chatham Islands stopped competing with one another and instead agreed to limit their catch and invest in research, forming the Chatham Islands' Shellfish Reseeding Association to enhance the production of paua.[23] Similarly, quota holders in the orange roughy, scallop, rock lobster, and snapper fisheries are investing in various research efforts to enhance fish stocks.[24]

Like New Zealand, Iceland has enjoyed success with ITQs as its main approach to managing fisheries. For example, the herring fishery was suffering from too many vessels and too few fish prior to ITQs. With introduction of ITQs, the fishery has improved markedly. Stock biomass is greater than at any time since the 1950s, and economic performance is vastly improved. In 1980, the first full year of ITQs, over 200 vessels took part in the fishery. Fifteen years later, in 1995, there were less than 30 vessels in the fishery. Yet the fewer vessels were harvesting twice as much herring as the 200 had in 1980. Productivity, as measured by catch per unit effort, has increased by a factor of five. There is also strong evidence of higher efficiencies in the capelin (a smelt-like fish) fishery. Since introduction of ITQs, total tonnage of the fleet has been reduced by 25 percent and fishing effort has contracted, with total days at sea reduced by 25 percent.[25]

A number of other countries are using various forms of ITQs in selected fisheries with success. For example, individual vessel quotas[26] in the British Columbia halibut fishery have helped spread out landings over a nine-month period, compared with a mere six days prior to their introduction. As a result, fishers supply the market with fresh halibut for a much longer period during the year, and their catch commands a much higher price. Further, the number of active vessels has decreased, reducing fleet crew payments and fixed costs.[27]

Australia's national government is using the ITQ system for the southern bluefin tuna fishery. Prior to ITQs, severe cutbacks in the total allowable catch and fleet capacity were required to sustain the fishery. After only six months under ITQs, the fleet capacity in the fishery had been reduced by 60 percent, as those who intended to stay in the fishery bought quotas from those who could earn more by leaving. As an indication of increased value of the fishery, ITQs began selling for just under $1,000 per ton on October 1, 1984; they sold for $2,000 per ton five and a half months later.[28]

Prior to ITQs, Greenland's offshore fishery in shrimp was largely open access. As a result, the fishing fleet expanded exponentially and individual catches and profitability became progressively smaller. Following introduction of ITQs, fleet size contracted and individual catches and profitability rose.[29]

To date, ITQs have been adopted in four federally managed fisheries in the United States.[30] The longest-running one is the Atlantic surf clam fishery. Prior to ITQs, the fishery had been suffering from overcapacity and long periods of downtime for crew and equipment. Allowable fishing time had to be steadily shortened due to increases in harvesting power. By the end of the 1980s, a surf clam vessel was allowed to fish only six hours every other week during the season. Under ITQs, the number of active vessels went from 128 in 1990 to 50 in 1997, thereby reducing excess capacity in the fishery. Those in the fishery were making better use of their resources. Fishing hours per surf clam vessel, for example, went from 404 in 1988 to 1,400 in 1994. In addition, ITQs improved vessel productivity to record levels. After two years under ITQs, catch per vessel almost doubled to 47,656 bushels.[31]

The Alaskan halibut fishery is a more recent example of ITQs in the United States. High fish mortality from abandoned gear, declining product quality, and hazardous fishing are some of the reasons ITQs were implemented in 1995. Following the introduction of ITQs, the length of the fishing season went from an average of 2-3 days per year between 1980 and 1994 to an average of 245 days per year under the ITQs. The longer season has meant that fresh fish is supplied to fish markets for a longer period during the year, thus product quality is improved. The longer season also allows fishers to fish during periods of good weather, thereby reducing hazardous fishing. The annual number of search and rescue missions for halibut fishers reported by the U.S. Coast Guard decreased significantly following introduction of ITQs. In addition, ITQs are helping conserve the resource. Managers report that unwanted fish mortality due to lost or abandoned gear went from 554.7 metric tons in 1994 to 125.9 metric tons in 1995, the first year of ITQs.[32]

Despite their growing acceptance, ITQs are not without criticism. Parzival Copes argues that ITQs can experience problems in bycatch fisheries.[33] These are fisheries in which the harvest of one species results in the harvest of another species. In New Zealand, ITQs are used in multispecies fisheries, and lessons learned indicate that this form of management can work if sufficient flexibility exists for balancing catches after the fact by acquiring additional quota holdings for bycaught species. Still, matching the mix of quota held to catches is problematic, and excessive bycatch has occurred in certain New Zealand fisheries. However, fishers appear to be making adjustments in their operations recently so that fewer overruns are occurring.[34] In addition, Copes argues that ITQs will not solve "high-grading," the tendency of fishers to discard smaller fish in hopes of catching larger, more valuable ones. High-grading appears to be a problem in some ITQ fisheries, but there is theoretical evidence that its occurrence depends on the conditions of each fishery.[35] Also, in certain fisheries, such as a lobster fishery where mortality from fish discarding is low, high-grading can be beneficial to the long-term health of the fishery.

There are two other potential problems with implementing ITQs that can negate their benefits. First, given that ITQs are generally set by a bureaucratic regime, there is the question of what incentive or ability bureaucratic managers have to establish the efficient level of harvest. The standard is to approximate the maximum sustainable yield. As discussed previously, this is usually not the yield that maximizes profits. To the extent that fishers can carry quota over into the next season or catch more in the current season by borrowing against future quota, they may adjust toward the economic optimum, but these options may not be available because of restrictions on quota transfers. In addition, even when sufficient scientific information on the condition of the stock is available to set a maximum sustainable harvest, political pressures can cause managers to ignore such information and instead allow unsustainable harvests. If the total catch is set too high, an ITQ fishery will suffer stock depletion.[36]

The second potential problem with ITQs stems from the time and money that would-be quota holders are willing to invest in order to secure claims to valuable quota rights. In other words, the race to catch fish will be replaced by a race for the quota.[37] If the quotas are allocated at random, by auction, or on the basis of historical catch prior to anticipation of ITQs, such investment will be minimized. However, if quotas are allocated on the basis of historical catch that anticipates the allocation, fishers will overfish in an effort to increase their share of the quota. Alternatively, if bureaucratic discretion determines the allocation, fishers will invest in influencing the decision. In either case, at least some of the potential profits from the ITQ fishery will be dissipated through the race for the property rights.[38]

FULL PROPERTY RIGHTS

Where feasible, a superior approach to either regulation or a politically managed ITQ system is to allow the establishment of full property rights. Exclusive rights to fishing areas are certainly not new. Robert Higgs found that Indians along the Columbia River had well-established rights to salmon fishing sites long before whites came to the area. "In some cases, these rights resided in the tribe as a whole; in other cases, in families or individuals."[39] The Indians developed effective technologies for catching the salmon and avoided overexploitation by allowing sufficient upstream migration for spawning purposes. Their "conscious regulation of the fishery played an important role in maintaining its yield over time."[40] Unfortunately, "legally induced technical regress" resulted from Washington State allowing interception of salmon at sea by whites and legislation that outlawed traps and effectively eliminated Indian fishing rights. This legislation ran counter to British common law, which had a place for private rights to coastal fisheries:

> . . . when we consider that there were already, in 1200 AD, in tidal waters, territorial fishing rights in England and a form of territorial salmon rights throughout the world in the 19th century, the legislative process can only be said to have reduced the characteristics of individual fishing right.[41]

For fish that are not mobile over wide ranges, property rights can be defined by specifying ownership of ocean surface area or of ocean floor, so-called territorial use rights in fisheries (TURFs).[42] Oyster fisheries along U.S. coasts offer a useful contrast of how property rights can improve resource allocation. Using data from oyster fisheries in Maryland, Virginia, Louisiana, and Mississippi from 1945 to 1970, economists Agnello and Donnelley tested the hypotheses that private ownership of oyster beds would generate more conservation and higher returns for fishers than open-access beds.[43] Under open access, we would expect fishers to take as many oysters as early as possible, with the result being diminishing returns later in the season. Agnello and Donnelley found that the ratio of harvest during the earlier part of the season to that of the later part was 1.35 for open-access oyster beds and 1.01 for private beds. After controlling for other variables, they also found that fishers in the private leasing state of Louisiana earned $3,207 per year, while their counterparts in the open-access state of Mississippi earned $807 per year. These findings support the expectation that private property rights solve the open-access problem.

The same property rights solution has been applied to other local fisheries. Some parts of Maine's lobster fishery continue to offer an example of private, community control of access. Anthropologist James Acheson describes the 100-year-old territorial system as a system "under community control" and at the same time "owned by the State." In order to harvest lobsters in a particular territory, fishers must be members of a "harbor gang." Nonmembers attempting to harvest lobsters are usually sanctioned by these extralegal harbor gangs. Though the system is not officially recognized by the state of Maine, it is recognized by Maine lobster fishers and has a significantly positive impact on productivity.[44]

Like parts of Maine's lobster fishery, a number of fishing communities around the world have avoided overexploiting fish stocks for decades by limiting entry and managing the use of coastal fishing grounds.[45] A large network of Fishing Cooperative Associations (FCAs) governing much of Japan's nearshore fisheries provides an example of a government-sanctioned community system. By law, FCAs own the fishing rights to specific territories extending as far as five and a half miles seaward.[46] The community approach, however, is not costless to maintain, and those that are not recognized by governments can easily succumb to governmental interference. For example, after decades during which local fishers successfully managed a coastal fishery in Valensa, Brazil, the Brazilian government decided to modernize fishing equipment by making nylon nets available to anyone who qualified for a bank loan arranged by the government.

The local fishers who had been managing the fishery did not qualify for the loan and did not have enough capital to purchase the nets on their own. Unfortunately, they were displaced by a few outsiders hired to fish the Valensa fishing grounds with the nylon nets. The local management system crumbled as old and new fishers fought over fishing spots, and eventually the fishery was overharvested and, ultimately, abandoned commercially.[47]

Economists Johnson and Libecap observed that even though private territorial rights in the Gulf Coast shrimp fishery were not formally recognized by federal and state governments, fishers historically resorted to informal contracting and the use of unions and trade associations to mitigate open-access conditions.[48] Fishing unions were particularly active from the 1930s through the 1950s, implementing policies along the Gulf Coast to limit entry, conserve shrimp stocks, and increase members' incomes. Such efforts by unions and fishing associations eventually met their demise in the courts, which refused to exempt the collective actions of associations and unions from antitrust prosecution:

> A cooperative association of boat owners is not freed from the restrictive provisions of the Sherman Antitrust act . . . because it professes, in the interest of the conservation of important food fish, to regulate the price and the manner of taking fish unauthorized by legislation and uncontrolled by proper authority.[49]

Although scarcity and competition limit the effectiveness of unions and associations, those organizations can provide an alternative for limiting entry and negotiating price agreements with wholesalers and canneries.[50] For a short time, they succeeded in internalizing the cost of regulations and conserving shrimp stocks. But as this situation revealed, any agreement establishing property rights to resources is difficult to maintain if the government declares it illegal.

The tradition of property rights is gaining popularity with the dramatic growth of aquaculture. From 1984 to 1995, world aquaculture grew from 6.5 million metric tons to 21 million metric tons.[51] With aquaculture, there is potential for increasing fish production while reducing pressure on wild stocks, though, as discussed below, there may be problems from pollution and escapement. Because investment in aquaculture requires secure property rights and because property rights are more likely to evolve where the costs of establishing rights are lower, sessile species, such as oysters, have the most promise. As noted earlier, private rights to oyster beds in some states have led to greater productivity.

The emergence of salmon ranching indicates that a solution based on property rights can also be applied to anadromous species that return to their original spawning ground. But before salmon ranching can realize its full potential, property rights problems have to be worked out. For example, a ranch operator has control over his stock only while the salmon are in captivity—before they are released and after they return for spawning. Otherwise, the salmon reside in the open sea beyond the rancher's control. Under conditions of the open range, the rancher may lose a substantial portion of his investment to natural mortality and commercial and sport fishers. Some of these problems can be overcome if there is better coordination between ranchers and commercial fishers. For example, because "restrictive ocean fishing season, depleted stocks of other species and low public smolt release levels raise the profitability from private aquaculture," economists Anderson and Wilen conclude that salmon ranchers would be willing to pay for a reduced season length and for reduced public smolt releases in return for receiving compensation from those who catch ranch fish in the ocean.[52]

This problem is similar to the one faced by British salmon sportfishers. In many cases, the fishers own fishing rights on the streams, but they are disturbed by the alarmingly depleted salmon stocks that return to spawn. The reductions are the result of increased commercial harvests, especially by fishers who have netting rights at the mouths of rivers.[53] To combat the problem, the Atlantic Salmon Conservation Trust of Scotland, a nonprofit group, has purchased 280 netting rights at a cost of $2.1 million and expects to reduce the netting catch of salmon by 25 percent.[54] Most have been purchased from private owners, but some have even come from the Crown. The idea of this buyout program began in Canada, where the federal government bought and retired commercial netting rights in New Brunswick, Nova Scotia, and Quebec.

Wild Atlantic salmon recovery was given another boost with temporary purchases of ocean salmon netting rights in Greenland and Faeroese fisheries. In 1989, Icelander Orri Vigfússon developed a proposal to buy commercial salmon netting rights held by fishers in Greenland and the Faeroe Islands. The Atlantic Salmon Federation fully supported Vigfússon's proposal and began raising funds immediately. In 1991, Faeroe Islands fishers agreed not to exercise their netting rights in 1991, 1992, and 1993. In return, they received $685,500 not to fish for salmon. As a result, in 1993, nearly twice as many salmon returned to their native rivers in Iceland and Europe.[55] In the same year, Vigfússon temporarily bought out netting rights of fishers in Greenland for 1993 and 1994. The buyout paid fishers $400,000 each year and reduced salmon netting off Greenland from 213 metric tons to 12 metric tons.[56] Such arrangements, which Vigfússon and the federation hope to make permanent through a comprehensive financing and job training program, are possible because the legal environment has established a system of transferable netting rights over which Vigfússon and fishers can contract.

Another approach that eliminates interactions between fish ranchers and fishers is based on raising salmon in pens. When the salmon reside in pens their entire lives, there are no losses due to commercial and sportfishing in the open ocean. This method has proven highly successful for the Norwegians, who are the leading international producers of Atlantic salmon. Salmon farmers and ranchers in the United States still face political opposition from commercial fishers, who have sought government protection of their markets, and from environmentalists, who fear that salmon farming will lead to more pollution of bays and inland waters. In 1987, commercial fishers in the Pacific Northwest convinced Alaskan state authorities to impose a one-year moratorium on net-pen salmon farms, and protests from local environmentalists in Washington have led the state to impose stringent guidelines in siting salmon farms.[57] More recently, pen-reared

Atlantic salmon operations in Maine have encountered opposition from seaside residents who do not want salmon farms spoiling their viewsheds and from regulators who perceive domestically raised salmon as a threat to wild salmon stocks.[58] In response, aquacultural interests cite scientific evidence that shows that salmon farming meets state water quality standards and poses little threat to wild salmon stocks.[59] In addition, they contend that the science used in salmon farming could be used to encourage survivability of strains of wild salmon.[60]

Institutional roadblocks also stand in the way of other private operations. In Maryland, out of some 9,000 acres of privately leased oyster grounds, only 1,000 are in production; 280,000 acres remain public. Privatizing Maryland's Chesapeake Bay oyster fishery faces the problem of weak enforcement of private leases. "It's hard to find an oyster ground that hasn't been poached upon," complained a planter on the Tred Avon River. "It's the main reason why so many people are reluctant to take their ground and invest their money in it."[61] Obviously, to realize the full potential of aquaculture in the United States, institutional barriers must be removed and the defense of private property rights must be strengthened.

Japan has led the way in establishing private property rights to fisheries. The Japanese took bold steps to allow the privatization of the commons because access to foreign fishing grounds was being more and more restricted by legislation like the Magnuson Act. The Japanese government now initiates the property rights process by designating areas that are eligible for aquaculture. The Fishing Cooperative Associations are then given the responsibility of partitioning these areas and assigning them to individual fishers for their exclusive use.

An exclusive right to harvest resources from a marine area allows an entrepreneur to invest in improvements and to capture the benefits of his or her investment. Consider the story of Ocean Farming, Inc., a company in the business of fertilizing the seas to enhance growth of phytoplankton, which in turn nourishes fish production. Based on actual experiments, company president Michael Markels estimates that with continuous fertilization, about one thousand tons of catchable fish per square mile can be produced each year. Recently, Ocean Farming entered into a contract with the Republic of the Marshall Islands giving the company an option for exclusive fishing rights on up to 800,000 square miles of deep ocean. Once harvesting starts, Ocean Farming will pay RMI $3.75 per square mile of ocean optioned or 7 percent of the value of the catch, whichever is more. Ocean Farming can charge other companies to fish the waters, and the firm has agreed to allow previous small-scale fishing to continue.[62]

This same approach could be used for many other fisheries that lend themselves to satisfactory control of resources within a given area. For example, suppose the National Marine Fisheries Service allowed people to "homestead" sections of the ocean within the 200-mile limit and harvest bottom fish from their homestead. One company, Artificial Reefs, Inc., recently completed a multifaceted artificial reef structure off the waters of northwestern Florida to enhance recreational fishing and provide an area for skin diving. The project was financed with grants, but it could have easily been financed by a potential owner of the area where the reef was deployed.[63] Experiments with sinking oil drilling platforms have demonstrated the success of this approach in improving fish stocks, so it is but a small step to encourage such investments through property rights. Owners of such homesteaded areas could catch the fish themselves, lease out fishing rights, or even lease recreational fishing rights.

There is sufficient evidence that the costs of defining and enforcing property rights for nonmigratory fish are low enough that we should lower the institutional barriers to such homesteading. In the Gulf waters off Alabama and Florida, even a very limited sense of ownership has spawned private provision of artificial reefs. These two states allow private entities to create reefs out of certain permanent structures in parts of their territorial waters. The reefs are considered public property as soon as they hit the water, but the sense of ownership that comes from knowing the exact location of a reef has been enough to encourage private initiative.[64] Unfortunately, the tenuous nature of reef ownership limits the potential for more privately created reefs.

For highly migratory fish, one property-rights solution is to establish the ITQ as a private property right—not a privilege revocable by the government—and allow the transition of managing the fishery from the government to the quota holders themselves. Like shareholding in a public company, "[h]olding an ITQ will allow large numbers of fishers . . . to come together and cooperate" in regulating fishing and coordinating their fishing with other users.[65] The costs of obtaining stock information, making management decisions, and monitoring and enforcing decisions "need not be higher—and can be much lower—than when these services are performed in a uniform way by a government agency." If there are some fisheries where costs are higher, then, as is done in New Zealand, fishers can literally hire the government agency to perform the services.

The New Zealand ITQ management system appears to be moving ever closer to a real system of privately owned fisheries. For example, Challenger Scallop Enhancement Company Ltd., whose shareholders are the owners of scallop ITQs, manage the fishery through contracts that allow the company to collect money for research and monitoring and enforcement of daily catch limits. They have even contracted with fishers in other fleets to reduce adverse environmental impacts from other fisheries on the scallop fishery.[66]

CONCLUSION

While they are by no means the final solution, current ITQ systems are a positive step toward addressing the tragedy of the commons in ocean fisheries. Their immediate benefit cannot be denied from the standpoint of ending the race to catch fish. Furthermore, the more secure they are as property rights, the more compelling it is for fishers to take a longer view of the resource. As is the case in New Zealand, ITQs have stimulated cooperation among fishers in enhancing fish stocks. It appears they make it "easier and cheaper" for fishers holding quotas "to act collectively" in managing the size of their catch.[67]

Unfortunately, the success of ITQs and collective action by fishing associations or communities stimulate political action by special interest groups that want a share of the growing pie. In countries where ITQs have become prevalent, some argue that the income generated by ITQs is a windfall to quota holders and belongs to the government.[68] Unfortunately, these conflicts focus only on how to carve up the pie, not how to make it even larger.

In addition, while ITQs offer considerable advantages, determining the size of the total catch remains a governmental function. Fishery regulators determine total catch based on biological sustainability and economic factors. Unfortunately, regulators are susceptible to political pressures from the special interest groups they regulate, who then become important bureaucratic constituents in the budgetary process. The incentive for pleasing such groups by maintaining an inefficient industry size can be strong enough to overshadow the objectives of efficient production and sustainable future catches.

A superior solution for species with limited range is to establish private property rights to their territories. Aquaculture in coastal areas is one example. There is also evidence that if the barriers to property rights to areas farther seaward are removed, entrepreneurs will invest in homesteading the oceans. In addition, we tend to overlook the extent to which communities hold de facto property rights to coastal fisheries around the world. While many are at risk from government interference, occasionally some arrangements are afforded legal recognition.

Establishing property rights will not be easy for highly migratory species, but, as in the frontier West, we can expect increasing efforts at definition and enforcement. Notably, technologies already exist that can facilitate defining and enforcing property rights in ocean fisheries. For example, transmitters on manatees use satellite telemetry to provide exact location, individual identity, water temperature, and the direction the individual is headed. "Devices can also be placed on board a fishing vessel to constantly relay its exact location via satellite, to identify whether it belongs in a certain area. . . . Heat-sensitive satellites cannot only monitor a ship's location, but can also use its heat profile to tell if it is towing nets or not."[69] In addition, a team of scientists in Kailua, Hawaii, have developed a same-day DNA field test to monitor whale stocks in the wild, and Norway is developing a DNA register on minke whale stocks.[70] For tracking individual whales, most promising are various tags that can be attached to a whale by firing a tag-carrying dart into its blubber. Each tag is actually a data-collecting and broadcasting unit capable of transmitting radio or sound waves.[71]

In the meantime, there are a number of steps that can be taken immediately to improve ocean fisheries. They include: implementing ITQ systems as property rights, removing legal roadblocks to managing fisheries collectively, providing legal recognition of territorial fishing rights, and refraining from further governmental redistribution of fishing rights. If carried out, such steps will move us a long way toward a free market environmental solution to the ocean commons problem.

CHAPTER 10

MARKETING GARBAGE:

The Solution to Pollution

"It may be garbage to you, but it's our bread and butter," reads the sign on the side of a garbage truck. This reminder provides a great deal of insight into the free market environmental approach to pollution. In the first place, garbage represents a cost that does not result in a valuable output. Wasted fossil fuel, for example, means that heat energy has been produced but not transformed into useful mechanical energy. Wasted wood means that a tree has been cut into useful pieces such as dimension lumber or furniture, but that 100 percent of the tree has not been transformed into these useful pieces. In the second place, garbage or waste to one party becomes another's "bread and butter" when the creator of the garbage must pay for its disposal. This brings us back to one of the central points of free market environmentalism, namely that there are competing uses for disposal space on the land or in the air and water. When there are property rights to the disposal space, the creator of the waste must pay the opportunity cost of using the space for which there are alternative uses. Therefore, in the case of a landfill, garbage and housing compete for the use of the land, and the garbage becomes the "bread and butter" of the landowner as long as the garbage producers must pay for the alternative uses that are forgone. Thinking about garbage as waste and as a competing use for disposal space helps us develop property rights approaches to garbage and pollution problems.

In this chapter we shall consider both the cost of waste and the liability for disposal as parts of the incentive equation that induces people to take responsibility for pollution and garbage. By carefully considering the roles of input prices and entrepreneurship in production decisions, we shall see how and why firms reduce waste and eliminate pollution. By considering the institutional arrangements that hold people accountable for their waste and pollution, we shall see how liability creates an incentive to take responsibility for waste.

ECO-EFFICIENCY

In the 1960s, few people could match the strength of the real muscle man who could crush a beer can with one hand, but today almost anyone can accomplish this feat. Is it because people have gotten stronger over the years? Of course not. It is because beer and soda cans are made of aluminum instead of steel, and the thickness of the aluminum has been drastically reduced since it was first used. What accounts for this transformation? The answer is that aluminum cans are much cheaper to produce and transport than steel cans, and there is always an incentive for companies to conserve on the costs of inputs and transportation. In 1963, it took 54.8 pounds of metal to produce 1,000 aluminum cans, but by the 1990s it took only 33 pounds per 1,000 cans. As a result, Anheuser-Busch was able to save 200 million pounds of aluminum per year.[1] Moreover, today's aluminum cans are 27 percent lighter than they were in the 1960s and 80 percent lighter than state-of-the art steel cans.[2]

Other inputs have similar conservation stories driven by economics. Steel used in high-rise buildings is 35 percent less than that used two decades ago. Fiber-optic cables made from 60 pounds of sand can carry many times more information than cables made from 2,000 pounds of copper.[3] Coal replaced wood as a heating fuel as forests were depleted. The transition in home heating from wood to coal to natural gas has reduced residential emissions of sulfur dioxide by 75 percent.[4] When the energy crisis first hit in the 1970s, people gave little thought to automobile fuel economy. But as real energy prices rose, people began thinking about miles per gallon and weighed the tradeoffs between size and safety and fuel efficiency. As real prices resumed their downward tread in the 1990s, consumers demonstrated the law of demand and switched back to larger, safer vehicles known as sport utility vehicles (SUVs).

In the case of wood products, dramatic changes have occurred in the amount of a log that is converted to useful products. One of the most significant sources of air pollution in lumber towns in the 1960s and 1970s was smoke from the tipi burner, a cone-shaped incinerator at the lumber mill where sawdust and wood chips were burned. Today tipi burners are rusty relics of the past not so much because of air pollution regulations, but because wood has gotten valuable enough to warrant conservation. Now trainloads of wood chips are transported from mills to production facilities where they are made into particle board. New technologies have produced better glues and ways of aligning wood chip fibers to produce a stronger board. The computer revolution has also found its way into the mill. Lasers scan logs as they enter the mill, and then computers determine the optimal lumber products to cut from each log and set the saws to make the cuts. As a result of this type of technology, wood losses fell from 25 percent in 1970 to less than 2 percent in 1993.[5]

Many firms are systematically thinking about how they can save money and help produce a cleaner environment through a concept known as eco-efficiency. "The first word of the concept encompasses both *eco*logical and *eco*nomic resources—the second says we have to make optimal use of both. . . . Reducing waste and pollution, and using fewer energy and raw material resources, is obviously good for the environment. And making better use of inputs translates into bottom-line benefits."[6] Following this concept, Minnesota Mining and Manufacturing (3M) implemented its "pollution prevention pays" program in 1975. In the two decades that followed, 3M reduced air pollution by 234,000 tons, water pollutants by 31,000 tons, wastewater by 3.7 billion gallons, sludge and solid waste by 474,000 tons, and energy consumption by 58 percent per unit of production. For the bottom line, this has meant $750 million in reduced production costs.[7] The Swiss chemical giant Ciba embarked on an eco-efficiency program that reduced energy consumption by 7 percent, water consumption by 17 percent, carbon dioxide emissions by 23 percent, sulfur dioxide emissions by 69 percent, volatile organic compounds by 42 percent, and chlorofluorocarbons per ton of product by 62 percent between 1991 and 1995.[8]

Though many environmentalists would argue that reductions like those achieved by companies such as 3M and Ciba should make sense regardless of cost, free market environmentalism (and eco-efficiency) emphasizes that such environmental gains are sustainable only if they pass economic muster. In the case of recycling, this means that the value of the resources saved must exceed the value of resources used to recycle. For example, recycling newsprint may reduce the number of trees that must be cut down for paper pulp, but recycling the newsprint is not free. It must be stored and transported to the recycling center, and from there it must be transported to a paper mill that can convert the fibers into useable paper, and part of the process requires using chemicals to remove the ink from the paper. After all of these costs are taken into account, we must ask whether recycling the paper was really worth it, especially given that the trees for pulp are a renewable resource. Bruce Van Voorst, writing for *Time* magazine, discovered that "more than 10,000 tons of old newspapers have piled up in the waterfront warehouses in New Jersey," and for the entire country this figure could exceed 100 million tons.[9] The fact that recycling centers cannot get rid of all the glass and paper that citizens dutifully bring them and must store it or dispose of it in landfills suggests that recycling is often not eco-efficient.

This does not mean that no recycling makes sense. In some locations, where waste disposal costs are high, where collection is easily centralized, and where markets for recycled materials are strong, recycling can represent a cost-effective way of handling discarded materials. Moreover, recycling materials such as aluminum that are easily isolated from the waste stream, that are easily transported, and that are valuable to manufacturers in the form of reduced energy, transport, and material costs, represents a cost-effective, efficient use of resources.[10]

Entrepreneurs have an incentive to find ways of increasing the value of recycled materials or of reducing the cost of recycling because they can profit from their entrepreneurship. For example, one environmental entrepreneur is making plastic lumber from recycled one-liter soft drink bottles. U.S. Plastic Lumber Corporation uses recycled plastic to make lumber for decks, boardwalks, piers, and even railroad ties. Though the plastic lumber costs more than the real thing, it requires much less maintenance and appeals to the environmentally conscious. With annual revenues in 1999 of $38.5 million, entrepreneur Mark Alsentzer says, "We're not tree huggers. But we are environmentalists by taking something that had no value and using it to replace wood."[11]

All of these examples illustrate the important role that input prices and entrepreneurship play in reducing waste. When input prices are low, there is little incentive to expend effort to try to conserve on the cheap inputs. Why worry about how to make the internal combustion engine more fuel efficient if fossil fuel is cheap? Why worry about the thickness of an aluminum can if it costs little to transport that can across the country? Higher resource prices create an incentive for entrepreneurs to find profitable ways to recycle. When input prices rise, entrepreneurs seek economically expedient ways to conserve resources. That is why study after study in the tradition of Julian Simon[12] has found little evidence of growing resource scarcity. When resources do begin to become scarce, people find substitutes either in the form of technologies that conserve on the use of scarcer resources or in the form of alternative resources that can do the job for a lower price.

DISPOSING OF WASTE

When waste cannot be reduced, the question is how to dispose of it without imposing unwanted costs on third parties. Returning to the central theme of free market environmentalism, the problem with waste disposal is that it creates a competing use for disposal media. As long as physical limits on production prevent us from converting all inputs into useable outputs, there will be waste and a demand for disposal space. The problem then becomes how to resolve conflicts over how disposal media should be used. The tipi burner described above was the method used by lumber mills to dispose of sawdust and wood chips. When they burned these unuseable inputs, smoke was put into the air. One tipi burner in wide-open spaces may not have created competition for the use of air because the air was not scarce, but when several tipi burners were located in a confined mountain valley, competition for the use of air between those wanting to dispose of their waste and those wanting a view was unavoidable.

There are three policy approaches to the disposal problem: regulation, tradeable permits, and property rights. The typical way of dealing with competing uses for air is to regulate when, where, and how much waste can be dumped into the air. Another possibility being increasingly used is to establish a specific amount of waste that can be dumped into the air and issue permits for that amount. Under this scheme, permit holders would decide whether to dump their waste into the air or whether to reduce their waste and sell their permits to others for whom it is too costly to reduce waste. Finally, the property rights approach establishes rights to clean air or rights to dump into the air and allows the holders of those rights to bargain over the optimal mix of competing uses, in this case clean air and garbage disposal. Let us consider these three policy approaches in more detail.

Regulation

Since the early 1970s, massive federal regulations in the Clean Air Act, the Clean Water Act, the Comprehensive Environmental Response, Compensation, and

Liability Act—better known as Superfund—and others have been the dominant approach to waste disposal. Under these laws, acceptable levels of discharge and pollution abatement are chosen and enforced by regulators who are not immune to politics. Fines may be levied on those who do not conform to the regulations, but the money collected seldom compensates those who are harmed.[13] Billions of dollars of private and public funds are spent on pollution control with little or no consideration for the cost-effectiveness of expenditures.

The success record of these regulations is less than stellar on almost every dimension. First, considering the trends in pollution abatement before and after the national government became the dominant regulatory force, it seems that national regulations have had less impact than commonly believed. Indur Goklany has compiled an extensive database that examines what we know about levels of air pollution in the United States during this century.[14] Considering a number of different pollutants, including dustfall, sulfur dioxide, carbon monoxide, ozone, nitrogen dioxide, and volatile organic compounds, Goklany concludes that "the nation's air is far cleaner today than it has been in several decades, despite the fact that population, consumption, and economic output—according to many environmentalists, the culprits fundamentally responsible for environmental degradation—have never been higher."[15] But the main cause for the improvements is not national regulations.

> The air is cleaner because, with the various environmental transitions, prosperity and technology were transformed from being problems responsible for air pollution to being solutions responsible for its cleanup. . . . By the time air pollution control was nationalized in 1970, air quality had improved substantially, particularly in the areas with the worst problems and for the pollutants perceived to be responsible for most of the public health risk.[16]

Second, national regulations have not delivered all they promised. Consider the effectiveness of Superfund. When Superfund legislation was passed in 1980, it was supposed to be a short, swift program to cleanse the nation of dangerous hazardous waste sites like Love Canal. It was to cost at most a few billion dollars and to be paid for mainly by those whose pollution causes serious harms or risks. But it has been "the shortcut that failed."[17] A Government Accounting Office (GAO) study reports that between 1980 and June 30, 1999, only 176 sites out of 1,231 had been cleaned up and removed from the National Priority List.[18] One purpose of Superfund was to reduce the health risk, especially cancer risks, from hazardous waste sites, but the Environmental Protection Agency (EPA) does not necessarily focus on the most risky sites. A study of 130 sites by the AEI-Brookings Joint Center for Regulatory Studies shows that "target risk levels chosen by regulators are largely a function of political variable and risk-protection biases." In one case, the EPA insisted that a site be cleaned to the point that children could safely eat dirt for 245 days per year without increasing their risk of cancer.[19] Given such ridiculous regulations, the AEI-Brookings study estimates that the average cost per case of cancer averted at a site is $11.7

billion.[20] Another GAO study reports that the average length of time for cleanup of the Superfund sites finalized by 1996 was 10.6 years, several times longer than cleanup under state laws.[21]

A classic case of failed regulation was the amendment of the Clean Air Act in 1977. Concerned about the problem of acid rain, Congress added New Source Performance Standards to the Clean Air Act in 1977 following its tradition of requiring strict application of "best available technology" standards for new coal-fired generating plants. Instead of setting specific emissions standards and allowing plants to meet them by using cleaner, low-sulfur western coal, the new standards forced owners of generation facilities to install stack-gas scrubbers, which cost more to buy and to operate.[22]

The reductions in sulfur dioxide could have been achieved at a much lower cost by burning low-sulfur coal, but a "clean air–dirty coal" coalition made up of eastern coal producers and environmentalists lobbied for the technological fix. Robert Crandall described the impetus for the strange coalition:

> Eastern coal producers feared that a sensible environmental policy would lead electric utilities to buy increasing amounts of low-sulphur Western coal. Since much of the Appalachian and Midwestern coal is high in sulphur content, it would eventually lose market share to the cleaner Western coal. Requiring stack-gas scrubbers for all new plants, regardless of the sulphur content of the coal burned, would eliminate the incentive for Eastern and Midwestern utilities to import low-sulphur Western coal. Environmentalists, for some reason, have had a burning desire to require utilities to install scrubbers, even though alternative technologies may be substantially less costly in most cases.[23]

Because the EPA could not monitor the efficiency of the scrubbers, the result was high new-plant compliance costs and a reduction in the rate of replacement of older, dirty utility boilers.

Rather than rational environmental policy, the clean air–dirty coal policy became a mechanism for redistributing wealth from electricity consumers who paid higher rates, to eastern coal miners, who feared losing their jobs. According to one estimate, electricity production costs were projected to increase by $4.8 billion by 1995 as a result of the 1977 amendments, and revenues to coal producers were expected to increase by $245 million.

> Since the environmental benefits were, at best, nil, this results in a maximum ratio of dollars gained by "winners" to dollars lost by "losers" of .05. That is, electricity consumers paid approximately one dollar for each 5 cents received by relevant coal producers.[24]

Through the regulatory process, the "best available technology" was applied at a very high cost when the far less expensive and more effective clean-coal alternative could have been implemented.

This experience demonstrates the difficulty of eliminating politics from the regulatory process. Some have suggested that the problem can be eliminated with general legislation and more autonomy for regulatory agencies such as the EPA, but this approach assumes that experts in the agencies will not be influenced by the lobbying pressures of those groups that have a stake in the outcome.[25]

With so much interest at stake in jobs, investments, and environmental quality, interest groups are not likely to sit back and let experts do what they think is right. The pressure from the environmental lobby for quick and dramatic solutions is strengthened by lawyers and industry groups who benefit from regulation. In this regulatory marketplace, politicians and bureaucrats are ready and willing to meet the demands of special interests. In the case of coal-burning utilities, politicians avoided drawing the ire of producers whose profits are tied to the operation of existing plants by applying strict, new regulations to new plants only. The result was a slowdown in the phase-in of cleaner-burning power plants. They also avoided the ire of environmentalists by appearing to take a tough stand against pollution.

Bruce Yandle captures the essence of political environmentalism in his "bootleggers and Baptists" theory: "This theory tells us that there must be at least two quite different interest groups working in the same direction—'bootleggers' and 'Baptists.'"[26] The title for his theory comes from the southern United States, where "bootleggers" are people who want to sell illegal alcoholic beverages on Sunday, and "Baptists" are the religious interests who want Sunday liquor sales eliminated. Each obviously gains from the passage of Sunday closing laws that prevent the legal sale of liquor on Sundays, but the bootleggers need the Baptists to give legitimacy to their goal. Similarly, businesses, such as the grandfathered electrical generators and the eastern coal mines, that can gain from costly, if questionable, environmental regulations need environmentalists to take the moral high ground and make the case for regulations. As Yandle puts it, "The Baptists lower the costs of favor-seeking for the bootleggers."[27] This theory of political environmentalism helps explain costly, ineffective regulations, from the 1977 Clean Air Act amendments in the United States to the 1997 Kyoto Protocol in Japan.

Finally, the costs of all these regulations have been astronomical. As of 1997, environmental regulations in the United States cost between $150 billion and $300 billion annually.[28] The costs imposed on the economy by the EPA itself are an estimated $100 billion annually.[29] Between FY1996 and FY1999, the EPA's budget grew by $800 million, from $6.5 billion to $7.3 billion. In the case of Superfund, the expected $1.6 billion of outlays in FY2000 represent an increase of more than $200 million over FY1999 outlays.[30] And between 1970 and 1996, the number of people employed by the EPA increased from less than 5,000 to nearly 20,000. All of which leads law professor and former environmental activist David Schoenbrod to observe that

> EPA regulations are so lengthy because those who write them respond more to pressure from within the agency to enlarge and protect its power than to the public's need for clear, concise rules. The problem is not that the agency is oversolicitous to the environment; it is that it is oversolicitous to itself.[31]

Schoenbrod concludes that the national administrative state embodied in the EPA "is many times more expensive, in terms of both direct cost and the drag on the economy"[32] than less centralized approaches and that the EPA should be stripped of most of its powers.[33]

Tradeable Permits

Because regulation has not always been effective and because it has been so costly, policy makers have turned to tradeable permits to introduce marketlike incentives into pollution control decisions. Under this method, a pollution control agency issues a limited number of permits authorizing the discharge of a specific amount of pollutants. The number of permits determines the level of pollution. While the initial number of discharge permits is determined politically, this system encourages polluters to consider the lowest-cost methods of control. If a polluter with a permit can reduce his pollution more cheaply than another, he has an incentive to reduce his pollution and sell his permit to the latter, profiting by the difference between the cost of reducing pollution and the sale price of the permit.

Tradeable pollution permits allow much more flexibility than the current pollution control regime. For example, polluters may be allowed to increase pollution at some location where water quality is high in return for reducing pollution in an area where it is low. Trading permits also reduces total control costs. If a permit is held by a firm that is capable of reducing pollution at lower cost than other firms, then a high-cost firm could purchase that right to pollute from the low-cost firm. The low-cost firm would then reduce the pollution level it had previously been allowed to discharge and still make a profit from the sale of the permit. By the same token, if an environmental organization wishes to reduce emissions, it can buy up permits.

One of the earliest candidates for this approach was water pollution control in northern Wisconsin, where a system of tradeable pollution permits was proposed for the lower Fox River. The river flows from Lake Winnebago to Green Bay, Wisconsin, and is lined with ten pulp and paper mills and four municipalities discharging effluent. Even with full compliance of BPT[34] industrial treatment and secondary municipal treatment, the desired dissolved oxygen levels were not reached during the summer. A tradeable permit system was proposed based on a simulation study that demonstrated a significant difference in costs of pollution treatment among the various discharge sources. The study found that control costs would be 40 percent higher with the traditional BPT approach to pollution control and that annual cost savings realized from a permit system would be from $5.7 million to $6.8 million.

Despite the obvious efficiency gains from tradeable permits, it took some time before they were brought into the EPA regulatory mix. Richard Andrews describes the long path to our current emissions trading program:

> In 1976, [EPA] used creative interpretation of the Clean Air Act to allow new pollution sources to locate in nonattainment areas if they obtained "offsets" from existing facilities that reduced their emissions. This inter-

> pretation was validated in the 1977 amendments to the law. In 1979 EPA announced a "bubble" policy, which encouraged firms to create excess "emission reduction credits" that could be traded to other sources to meet their emission reduction requirements.[35]

With this foundation, the 1990 amendments to the Clean Air Act established total sulfur oxide emissions caps and allowed polluters to trade permits for sulfur oxide emissions nationwide. Andrews summarizes:

> By 1997 this approach had produced exceptionally promising results. In 1996 all 445 utility boilers and combustion turbines that were subject to these regulations met their required standards. Utilities had reduced their emissions 30 percent below what the cap required, at about half the expected cost, and acid rain in the Northeast had declined by an estimated 10 to 25 percent.[36]

Trading sulfur oxide was such a success because it allowed utilities to use more western low-sulfur coal, thus substituting control of total emissions for installation of specific technological fixes required in earlier Clean Air Act amendments. Unfortunately, emissions trading for other air pollution sources has been less successful because they are "still closely tied to the preexisting point-by-point, technology-driven control requirements, leaving only slim opportunities for additional benefits."[37]

Another successful version of a tradeable permit system to control point and nonpoint sources of pollution has been used to improve water quality in North Carolina's Tar-Pamlico Sound.[38] Sores on fish, algal blooms, and fish kills were the outward indicators that water quality in the sound was suffering from nitrogen and phosphorus nutrient loading. As a result, in 1989 the North Carolina Department of Environmental Management (DEM) declared the sound to be Nutrient Sensitive Waters and established the basis for regulation. With 85 percent of the nutrient loads originating from nonpoint agricultural sources, however, regulatory solutions were costly, if not ineffective. Publicly owned treatment works and industrial plants were already trying to reduce their discharges of nutrients, so that further reductions would have had little effect and would have been quite expensive.

As an alternative, a watershed market was proposed by the Tar-Pamlico Association, made up of the publicly owned treatment works and some industrial plants. Under the trading plan, the members of the association agreed to assess themselves a fee and to use the proceeds to study the nutrient loading problem and help fund efforts by farmers to implement best management practices that would reduce nonpoint sources of nutrient loading. Such management practices include raising the banks of sewage-collection ponds for confined hog feeding operations and spreading animal waste on lands where it will not leach into surface and groundwater. In return for helping reduce the nonpoint source loading, the point sources were given credit for reducing the overall nutrient loading in the sound. "Collectively, the non-point source dischargers would have to spend between $50 and $100 million to meet the tighter state standards. Under the trading plan, however, the estimated cost of reducing the same quantity of nutrients was (only) $11.7 million."[39]

Industrial effluent delivered to municipal treatment plants is another type of pollution that is amenable to a system of tradeable permits. As a means of controlling these discharges, the EPA has established a set of pretreatment standards. These standards share the deficiencies of all standards, namely a lack of cost-effectiveness, but tradeable discharge permits can be applied to overcome such deficiencies. For example, consider an electroplating operation in the Rhode Island jewelry industry that discharges high concentrations of cyanide, copper, nickel, and zinc into municipal sewer systems. Because the municipal treatment plants in the area are not capable of removing these substances, the EPA applies the pretreatment standards to the entire industry. If strictly applied, such standards would force 30 to 60 percent of the small firms out of business. One analysis of alternatives for meeting water quality objectives, however, concluded that pretreatment standards meet the objectives at a cost that is almost 50 percent greater than the tradeable permit alternative. An effluent permit system would achieve the target level of water quality at a cost of $12.5 million; the pretreatment-standard approach would achieve the target at a cost of $19.3 million.[40]

Property Rights

The free market environmental approach to pollution is to establish property rights to the pollution disposal medium and allow owners of those rights to bargain over how the resource will be used. Following the lesson from Ronald Coase,[41] the central questions are: who has the right to use the air, water, or land? for what uses? and what are the transaction costs for the owners of these rights to bargain over the uses of the resources? If the property rights are well defined and enforced and the bargaining costs are low, regulation and tradeable permits are unnecessary; a market of waste disposal will determine which of the competing uses dominates.

Of course, most people argue that the problem with pollution is that property rights are not well defined and that the bargaining costs are high. But are these presumptions accurate? Consider the case of household garbage destined for the landfill. The main concern with solid waste disposal is whether it will spill out of the disposal site and cause damage to other landowners. Property rights make it the responsibility of homeowners to hire a municipal or private company to haul the garbage to a landfill so that the garbage does not pollute the neighbors' property. And once the garbage is at the landfill, as long as the owner of the landfill is held accountable for any garbage that may flow from its boundaries, either by being blown by the wind or by seeping through the ground, no third parties will be harmed by the transaction. The garbage may not be pleasant to either the household or the landfill owner, but the transaction between the two transforms the garbage into "bread and butter." It is no longer garbage to the landfill owner because the creator of the waste must compensate the garbage hauler and the landfill owner, both of whom willingly accept responsibility for the garbage. Compensation gives the person who generates the garbage an incentive to compare the benefits of garbage production with the costs of disposal to arrive at an optimal level of both.

But are property rights to the waste and to the medium into which it is disposed always as clear as they are in this simple example? Especially when waste is dumped into the air and water, are the property rights clear so that producers and disposers of waste are accountable for the cost of their competing use of the waste disposal medium? As we shall see, the complexity of the problem varies with the medium into which the waste is dumped and with the number of parties involved in the competition for disposal space, but there is abundant evidence that the common law of property, nuisance, and torts is a way of making people accountable for their waste.

For example, why would a power-generating company go to great pains to prevent fly ash removed from its smokestacks from seeping into the groundwater? Indeed, in the 1970s and 1980s, the Montana Power Company built new coal-fired generating plants in remote parts of eastern Montana. Complying with EPA air pollution regulations, the company installed precipitators on its smokestacks to remove fly ash from the air and collect it in settling ponds. From these ponds, the fly ash could leak into the groundwater, creating a potential health hazard for humans and livestock, but even without any regulations the company went to great lengths to prevent this from happening. It installed a primary catchment barrier under the pond, a secondary barrier under the primary barrier, a tertiary barrier under the secondary barrier, and a sump pump under all three to collect any leakage and pump it back into the pond. Why would the company go to all this expense? In a word, liability. If contamination of groundwater occurred, it would be clear who was the owner of the waste, i.e., the power company, and who was the owner of the medium, i.e., the owner of the well. Hence, the precedence of common law made the property rights clear and provided an incentive for the company to safeguard against liability for causing harm to others.

Similarly, a lead smelter in Tacoma, Washington, took pains to reduce the impacts of acid rain long before that term was a household name among environmentalists. The smelting process generated sulfur dioxide emissions that could mix with moist air and generate acid rain. A tall stack allowed the smelter to mix its sulfur dioxide in the upper atmosphere and dispose of its waste without impacts on others. But when local atmospheric inversion occurred, emissions from the stack resulted in localized heavy concentrations of acid rain that caused damage to plants and homes near the smelter. When this occurred, the company quickly offered compensation to the harmed individuals because it knew it would be found responsible if lawsuits arose. Again, ownership of the waste, sulfur dioxide, and the medium, the vegetation, was clear. Following the logic of Coase, the company realized that the transaction costs were much lower if the company recognized the property rights of local landowners and paid for the damages it created. There was not costly litigation because the property rights were clear, and there was an incentive to find an optimal level of garbage generated because it had to pay for damages it caused.

The infamous story of Love Canal, one of the hazardous waste sites used to justify passage of the Superfund legislation, is another example of liability rules forcing a potential polluter to take account of its actions.[42] The common version of the story is that Hooker Chemical Company disposed of hazardous wastes at a site near Niagara Falls, New York, and that these hazardous wastes leaked into adjacent lands causing health problems. It is true that Hooker Chemical did dispose of hazardous waste at its privately owned Love Canal site, but it did so in a very responsible way in the late 1940s before Superfund regulations were even a glint in the eye of Washington regulators. The site was chosen because it was an abandoned canal, which meant that it was quite impervious to leakage. Then, before chemicals were placed in the canal, further efforts were taken to prevent the canal from leaking by lining it with clay, and all indications were that the precautions were effective. Like the Montana Power Company, Hooker Chemical did this because the common law made it clear that the company would be liable for any damages that resulted from chemicals leaking from the site.

So why did the leaks come about? In 1952 the local school board was searching for land on which to build a new school. Seeing unused land at the Hooker Chemical site, the board approached the company about buying the land for the school. One might think that this would give Hooker Chemical an excellent opportunity to get rid of the site, but to the contrary, the company refused to sell the land, realizing that liability might follow the original owner, and the company even went so far as to take school officials to the site to show them what was buried there. Nonetheless, the school board threatened condemnation of the land, so Hooker Chemical sold the land to the school district for $1, writing explicitly into the deed that the site contained hazardous wastes for which the new owner would be liable. The deed reads:

> Prior to the delivery of this instrument of conveyance (deed), the grantee (Niagara Falls Board of Education) herein has been advised by the grantor (Hooker Electrochemical Company) that the premises above described have been filled, in whole or in part, to the present grade level thereof with water products resulting from the manufacturing of chemicals by the grantor . . . and the grantee assumes all risk and liability incident to the use thereof. It is, therefore, understood and agreed that, as part of the consideration for this conveyance and as a condition thereof, no claim, suit, action or demand of any nature whatsoever shall ever be made by the grantee, its successors or assigns, against the grantor, its successors or assigns, for injury to person or persons, including death resulting therefrom, or loss of or damage to the property caused by, in connection with or by reason of the presence of said industrial wastes.[43]

Obviously the school could not be built without disturbing the integrity of the canal, and to make matters worse, part of the site was sold for a subdivision requiring sewer and water line ditches that broke through the canal walls. Not surprisingly, the site leaked. And when state health officials started investigating the toxicity of the site in 1978 and subsequently recommended that people be evacuated from the area, President Carter declared the site a national emergency.

When the issue hit the courts, the deed notwithstanding, Hooker (now owned by Occidental Chemical Corporation) was accused by the state of negligence. Though the judge did not feel there was sufficient evidence to support the full negligence allegation, the court did not approve of Hooker's conduct because it did not specifically show the school board that there were chemicals exactly where the school would be built and did not specifically mention how shallow the disposal barrels were. Without a finding of negligence, the company was not fined or penalized, but a settlement with EPA required that Occidental Chemical Corporation pay Superfund $102 million to reimburse incurred cleanup costs, $27 million to the Federal Emergency Management Agency, and $375 for wildlife damages.[44] Common law liability holding Hooker Chemical responsible for its original waste disposal actions seemed to work, and subsequent costs to Occidental Chemical certainly raise the cost of waste disposal, but legal process under Superfund has not held the school board, which "let the cat of the bag," liable for its actions.

The skeptic may respond that these are easy examples because they involve land and groundwater, to which property rights are easier to define and enforce, but we shall see below that common law has been dealing with pollution problems for a long time. However, in order for the common law to work, owners of property must be able to enforce their rights against trespass, and this requires that plaintiffs be able to prove who is creating the pollution and what the damages are that result from that pollution. Because the costs of identifying polluters and their damages are positive and sometimes quite high, the common law will not always result in well-defined and enforced property rights to disposal media.

But just as competition among cattlemen on the open range led to the invention and proliferation of barbed wire, technological change can lower the costs of defining and enforcing property rights to air and water. Fred Smith, president of the Competitive Enterprise Institute, let his imagination consider the possibilities for barbed-wire-type solutions to establishing property rights to air and water:

> Tracers (odorants, coloring agents, isotopes) might be added to pollutants to ensure the damages were detected early where the costs of reduction were lower. Detection and monitoring schemes would evolve as environmental values mounted and it became appropriate to expend more on fencing. There are exotic technologies that might well play a fencing role even for resources as complex as airsheds. For example, lasimetrics, a technology which can already map atmospheric chemical concentrations from orbit, might in time provide a sophisticated means of tracking transnational pollution flows. If that system were combined with a system under which each nation adopted some fingerprinting system to identify its major greenhouse gases (a type of chemical zip code system), it would become possible to trace pollution to its source and thus make it possible to make the polluters pay. Note that most developed nations do participate in "labeling" high explosives manufactured in their countries as part of the worldwide anti-terrorist program.[45]

The potential for these imaginative solutions to establish ownership of waste depends on contaminant source analysis, of which there are three basic types.[46] Compositional analysis involves the identification of specific components of a contaminant through molecular or isotopic composition. For example, compositional analysis can be used for identifying sources of oil contamination, since oil is a mixture of various molecular components and the fraction of each component varies with the source of the crude oil. Because lead, with its four stable, nonradioactive isotopes, is unique among all metals, it lends itself rather well to compositional analysis. In the case of *Ethyl Corporation et al. v. Environmental Protection Agency* (1976), the plaintiffs challenged the EPA's gasoline lead regulations on the grounds that lead regulations were arbitrary and capricious because they were not based on evidence that lead "will endanger" people. The EPA prevailed, however, with a clinical lead isotope study that distinguished between inhaled and ingested lead. By controlling the dietary intake of lead and using a lead isotope tracer, the clinical study was able to show that approximately 30 percent of blood lead was inhaled rather than ingested. While this study was used to support pollution regulation, it illustrates how compositional analysis can be used to identify sources of pollution, a step necessary in a property rights approach to pollution control.

The second type of contaminant source analysis uses tracers that can be discharged with the potential contaminant to identify the source. Tracer analysis was used in *Central Arizona Water Conservation District et al. v. United States Environmental Protection Agency* (1993), in which petitioners contended that regulations requiring a 90 percent reduction in sulfur dioxide emissions from the Navajo Generating Station were arbitrary and capricious.[47] To identify the source of the pollutants, the Winter Haze Intensive Tracer Experiment (WHITEX), sponsored by a consortium of government agencies and utility companies, established a battery of air monitoring stations in and around the Grand Canyon and injected deuterated methane, a sulfur tracer that mimics the dispersion behavior of sulfur dioxide but is not found in ambient air, into the stack of the 2,250-megawatt Navajo Generating Station (NGS) in Page, Arizona. The experiment concluded that sulfur dioxide from the plant was responsible for 70 to 80 percent of the sulfates contributing to the haze. Though a subsequent study, conducted by the operators of the generating plant using perfluorocarbon tracers, attributed a lower percent of responsibility to the generating plant, the EPA continues regulating the plant on the grounds that visibility impairment is traceable to NGS and that "NGS [was] a dominant contributor to certain visibility impairment episodes." Again, this example buttresses regulations rather than the enforcement of property rights, but it also illustrates the potential for identifying specific sources of pollution. Tracers could act like cattle brands or dog tags. Just as states register cattle brands and prosecute rustlers and as towns license dogs and record who their owners are so that they can be held accountable for the actions of the dogs, the government could move us in the direction of a property rights solution to pollution if it would require potential polluters to tag their emissions with tracers and monitor the flow of pollutants in the atmosphere.

The third method of contaminant source identification is contamination distribution, the analysis under contention in the novel and movie *A Civil Action*. By this method, the distribution of the contamination itself can be used to determine the contaminant's source. In *Anne Anderson et al. v. Cryovac, Inc., et al.* and *Anne Anderson et al. v. Grace Co. et al.,* the plaintiffs living near two city wells in Woburn, Massachusetts, alleged that solvents from a manufacturing plant owned by W. R. Grace & Company and from a 15-acre parcel of vacant wetland owned by the John J. Riley Company, then a subsidiary of Beatrice Foods Company, caused various ailments, including leukemia. The district court tried the cases in stages, first attempting to determine responsibility for the pollution and then determining the causation and damages. Contamination distribution analysis necessitates modeling the transport behavior of the contaminant and the medium through which it is traveling. In the Woburn cases, the judge ruled that the hydrological modeling studies were unable to prove that the Beatrice land was the source of contamination, but the jury did hold Grace liable for polluting the wells. An out-of-court settlement, however, eliminated the need for court determination of causation and damages.[48]

All three of these examples suggest that contaminant source analysis is difficult to apply conclusively, but they also suggest that science can help defend property rights to clean air and water against trespass from pollutants. As in the early years of barbed wire, more investments will be necessary to perfect contaminant source analysis, and this investment is more likely to be forthcoming in a legal environment that depends on property rights rather than regulations. For this reason, the common law provides a model for how free market environmentalism might deal with pollution. The Rhode Island case of *Wood v. Picillo*[49] reflects how common law courts can react to changes in the technology. The case questioned whether a farmer who maintained a hazardous waste site on his property was harming the neighbors with noxious fumes and ground and surface water pollution. A 1934 court had held that plaintiffs were required to show that the defendant could "foresee" the consequences of pollution and found for the defendant on the grounds that the science of hydology was too "indefinite and obscure" for such vision. The 1982 ruling in *Wood v. Picillo,* however, overturned the 1934 decision on the grounds that

> the science of groundwater hydrology as well as societal concern for environmental protection has developed dramatically. As a matter of scientific fact the course of subterranean waters are no longer obscure and mysterious. . . . We now hold that negligence is not a necessary element of a nuisance case involving contamination of public or private waters by pollutants percolating through the soil and traveling underground routes.

In other words, property rights can evolve with technology.

COMMON SENSE FROM THE COMMON LAW

In his discussion of "Coase, Pigou, and Environmental Rights," economist Bruce Yandle provides one of the best defenses of common law as an ally of the property rights approach to environmental quality.[50] Thinking of environmental problems as competing uses for resources, Yandle emphasizes that resolution of this competition is accomplished best by defining and enforcing property rights to the resource in question and by letting parties bargain over which uses will prevail. At common law, people have a right to have their property free of pollution, but they "must be able to demonstrate damages to obtain relief in court. Loose assertions about environmental quality and the need to protect it will not do the job. Ownership of damaged property or loss of recognized rights must be shown. Information relevant to the harm must be provided."[51] In other words, under the property rights approach to pollution, the plaintiff must demonstrate that there is a connection between the cause and the effect of pollution, that the defendant is responsible for the cause, and that damages have resulted.

It is precisely these elements of common law that create skepticism about common law as a way of protecting property rights against invasion from pollutants.[52] The argument is that the burden of proof under the common law creates unnecessarily high transaction costs and litigation, which in turn will make environmental improvements impossible. But careful examination of common law pollution cases shows that the judges were environmentalists long before regulators in national capitals. Elizabeth Brubaker, a scholar of Canadian common law, points out that nuisance, which is defined as "anything done to the hurt or annoyance of the lands, tenements, or hereditaments of another," provides the basis for environmental protection:

> People have used it to protect themselves from pesticide sprays, smoke soot, steam, dust, fumes, and other air pollutants. Road salt has been successfully challenged under nuisance law, as have leaking oil tanks and seeping privies. Foul smells are often found to be nuisances, as are noise and vibrations from commercial and industrial operations. In the 1920s, one judge went so far as to say, "Pollution is always unlawful and, in itself, constitutes a nuisance."[53]

There is a rich history of common law cases that illustrate how people have protected themselves from pollution prior to passage of federal pollution legislation in the 1970s. These cases tell us how common law has worked in the past and how it might still work if given a chance.[54] In *Carmichael v. City of Texarkana,*[55] for example, the Carmichaels owned a 45-acre farm in Texas on a stream that bordered Arkansas. The city of Texarkana, Arkansas, built a sewage system that collected city sewage and dumped it "immediately opposite plaintiffs' homestead, about eight feet from the state line, on the Arkansas side." As a result of the sewage, the Carmichaels were forced to obtain domestic and stock water from another source at a cost of $700. Believing they were harmed by the sewage, the Carmichaels sued the city seeking damages and injunctive relief. The court found that Texarkana's

> cesspool is a great nuisance because it fouls, pollutes, corrupts, contaminates, and poisons the water of [the creek], depositing the foul and offen-

> sive matter . . . in the bed of said creek on plaintiffs' land and homestead continuously . . . depriving them of the use and benefits of said creek running through their land and premises in a pure and natural state as it was before the creation of said cesspool.

Though the city was operating properly under state law, the judge in the case found that this was no excuse for fouling the water, noting that he had "failed to find a single well-considered case where the American courts have not granted relief under circumstances such as are alleged in this bill against the city." It is worth emphasizing that the Carmichaels prevailed in 1899, long before clean water regulations were being considered at any level of government.

Whalen v. Union Bag & Paper Co.[56] shows how a single farmer could stand up against a large company even when the industrial revolution was in full swing. In this case, Mr. Whalen owned and operated a farm downstream from a new pulp mill that cost $1 million to build and that used the creek to dispose of its waste. Whalen sued the mill seeking damages and an injunction, contending that its pollution made the water unfit for agriculture, and he prevailed. An appellate court reversed the injunction based on the company's argument that the original court did not take account of the value of the mill and the 500 jobs it created. New York's highest court, however, reinstated the injunction, stating:

> Although the damage to the plaintiff may be slight as compared with the defendant's expense of abating the condition, that is not a good reason for refusing an injunction. [This] would deprive the poor litigant of his little property by giving it to those already rich. . . . The fact that the appellant has expended a large sum of money in the construction of its plant, and that it conducts its business in a careful manner and without malice, can make no difference in its rights to the stream. Before locating the plant the owners were bound to know that every riparian proprietor is entitled to have the waters of the stream that washes his land come to it without obstruction, diversion, or corruption, subject only to the reasonable use of the water, by those similarly entitled, for such domestic purposes as are unseparable for and necessary for the free use of their land.

In other words, it was Whalen's clearly defined property rights that protected him from invasion and therefore required that the company bargain with Whalen and other riparian owners before competing with their uses of the stream.

People could also protect themselves from air pollution, as illustrated in *Georgia v. Tennessee Copper Co.*[57] Georgia filed a public nuisance action on behalf of its citizens against companies operating two copper smelters in Tennessee near the Georgia border. Georgia ultimately prevailed in the U.S. Supreme Court, with Justice Holmes finding that a public nuisance was created because the "sulphurous fumes cause and threaten damage on so considerable a scale to the forests and vegetable life, if not to the health, within [Georgia]."

In Utah, where mining was a very important industry, *Anderson v. American*

Smelting & Refining Co.[58] found in favor of farmer Anderson, who sought an injunction against one of the nation's largest smelters. He claimed that its arsenic and sulfur emissions were damaging his cattle and crops. In Utah's political setting, one might reasonably assume that the smelter would prevail, but not so. The company countered that there was no way to reduce emissions given the magnitude of its operations and that the court should consider the fact that the smelter was providing hundreds of jobs, as opposed to the harm to a single farmer. But the court found for Anderson even though "there can be no solution [to the] smelting smoke problem . . . unless SO_2 is removed entirely from the smoke stream . . . or so diffused . . . [that] the concentration . . . will be reduced to a point imperceptible to the senses." Moreover, the fact that the smelter was such a large employer did not matter; according to the court, no "industry, however important, can justly claim the right to live or operate which creates a nuisance in operation or trespasses upon the property or the inherent personal rights to others."

So if the common law is effective, as these few cases might suggest, why is it not the prevalent method of dealing with pollution? Yandle's answer is that regulation has pushed the property rights approach into the background. He believes two factors account for this transition:

> First, pollution cases involving more than one state, covered by what was called federal common law, became subject to federal statutes. Judges deferred to federal administrative bodies, like the U.S. Environmental Protection Agency, and to statutory remedies. . . . The environmental statutes offered another attractive feature for environmentalists who could not make a valid claim of damages in a common-law court, but who nonetheless felt driven to correct perceived environmental harms. Under the statutes, any citizen could file an administrative complaint, which merely had to show an infraction of rules created by statute, and seek access to a federal court.[59]

As an illustration of how statutory law weakened common law protection, Yandle offers the case of *Illinois v. Milwaukee.*[60] In this case, Illinois sued Milwaukee for its sewage discharges into Lake Michigan, claiming a public nuisance for Illinois residents obtaining drinking water from the lake. Believing that "a state with high water-quality standards may well ask that its strict standards be honored and that it not be compelled to lower itself to the more degrading standards of a neighbor," the U.S. Supreme Court enjoined Milwaukee from continuing its discharges.

Not long after the Supreme Court ruling, Congress passed the 1972 Clean Water Act, turning water quality over to the EPA and allowing it to establish the technical compliance rules. Because Milwaukee believed that it was in compliance with the Clean Water Act, it returned to court with *Milwaukee v. Illinois*[61] in 1981. In this case, the Supreme Court found for Milwaukee, holding that federal common law could not impose a standard more strict than the EPA standard, and therefore lifted the injunction.

This does not mean that common law is not still used as a defense against invasion by pollution. In *New York v. Schenectady Chemicals, Inc.,*[62] for example, the court dismissed a claim under a state environmental statute, relying instead on the common law rule of strict liability to defend individual property rights against invasion. Similarly, in *New York v. Monarch Chemicals et al.,*[63] the appellate court used liability based on nuisance to hold the landlord and his tenant accountable for pollution despite the defense that New York's Environmental Conservation Law did not apply to the problem.

Nonetheless, in the competition of property rights versus regulatory approaches to pollution problems, there can be little doubt that the regulatory approaches are winning. Data on the share of common law nuisance and trespass cases relative to all air and water pollution cases for 1940 to 1994 show a systematic and dramatic decline from over 50 percent of all cases to fewer than 20 percent.[64] Yandle believes that at least "part of the explanation must relate to the rise of state statutes and regulations."[65] Thus the question becomes whether the regulatory approach is better than the property rights approach.

Nor should these cases be taken to suggest that common law will always work. Meiners and Yandle acknowledge that transaction costs can be high, especially if there are multiple polluters and if injuries and harms come after a long gestation period. In this case, "Regulation may be the only answer. . . . But we cannot know how the law might have evolved had it not been pushed to one side by regulation."[66]

In some cases, it is public ownership that affects transaction costs. Automobile emissions, for example, are considered nonpoint source pollution because they involve multiple, mobile polluters. But for the most part, automobiles travel on public highways. Suppose instead that the roads were privately owned and that common law were applied. It would not be a stretch, given the above examples, to imagine the courts considering the highway the source of pollution and holding the owner of the highway accountable. The owners of the private highway would have an incentive to reduce emissions, so cars with better pollution control equipment might receive lower tolls, and those with no equipment might be banned altogether. Moreover, the freeway owner could earn higher profits by reducing congestion, increasing speeds, and thus lowering pollution that results from stop-and-go traffic. Laser technology is making it easier to measure tailpipe emissions, and bar-code sensors and cards make it easy to monitor highway users and charge differential prices.[67] Again, this is not to say that common law will always work, but it does suggest that we should not be too quick in assuming that the transaction costs for property rights approaches to pollution cannot be overcome.

CONCLUSION

For the most part, policy analysts proposing market approaches to pollution problems have focused on tradeable discharge permits. This approach certainly offers an effective way of introducing the discipline of the market into pollution abatement, but permits require political control and do not provide for a complete market in pollution. Even these schemes have been slow to evolve in the regulatory regime that has been in place since the 1970s.

A free market solution to the pollution problem would require well-defined and enforced property rights to both the waste and the disposal medium. Though establishing property rights to air and water is costly, we should not forget the importance of the evolution of property rights and the common law. As clean water and air have become more valuable, entrepreneurs have a greater incentive to define and enforce rights to the resources. If we continue to subsidize the use of these resources and to subsidize the costs of disposal, however, entrepreneurs will not be getting the right signals. At the very least, municipal, state, and federal agencies should raise the price to individuals for using waste disposal systems. Not only will this give people incentives to produce less garbage, it will also provide funds to clean up pollution.

At the same time we should reconsider the efficacy of common law as a solution to pollution problems. David Schoenbrod, a lawyer formerly with the Natural Resources Defense Council and a former advocate of the regulatory approach, asks, "Is real reform feasible?"[68] He believes it is possible if modern legislative practices would incorporate a few fundamental changes:

> First, the federal government should deal with only those environmental issues that the states are institutionally incompetent to handle. . . . [T]his would leave most pollution regulation at the state and the local levels. Second, legislation, whether at the federal, the state, or the local level, should be restricted to enacting rules of conduct, not ideals such as "to protect health" and "the public interest.". . . The consequence of these two changes would be that legislated rules of conduct would tend to have many of the virtues of the common law. . . . To complete my wish list for remaking statutory environmental law in the spirit of the common law, real injury would be required for standing; permits requirements would be eliminated. . . . ; remedies would emphasize, to the extent possible, damages and injunctions tailored to prevent harm, rather than emphasizing criminal and civil penalties; and such penalties would be barred when there is neither harm nor fault.[69]

Schoenbrod recognizes that this would not be a perfect world, but it would certainly move the pollution debate in the direction of property rights and markets.

CHAPTER 11

CALLING ON COMMUNITIES

In his influential 1968 article, "The Tragedy of the Commons," Garrett Hardin explained why a scarce resource open to all is subject to overexploitation.[1] He used as an example of the commons a pasture open to all herdsmen for cattle grazing. Hardin pointed out that eventually the pasture will become overgrazed. The reason? Each herdsman can capture all the benefits of adding more cows, while facing only a fraction of the costs—the harm caused by excessive grazing—because costs are shared by all. The tragedy, notes Hardin, is that each individual is "locked into a system" of competition for grass that leads to ruin.[2] A similar tragedy occurs when a fishing territory is open to all fishers. Each fisher captures all the benefits of harvesting more fish, while facing only a fraction of the costs—the reduction of the fish population for future harvest.[3] Similar logic can explain the deterioration of other resources, such as airsheds and waterways open to all for dumping purposes.

While Hardin's logic is useful for predicting outcomes when cooperation among users is not possible, it fails to account for those cases in which users have cooperated and coordinated their use of the commons. Recall that as the value of grazing rose on the western frontier, cattlemen formed associations that proved instrumental in restricting entry onto the range and pressuring state and territorial governments to pass laws to punish those who violated rules of use.[4] Cattlemen also found cooperation profitable when it came to joining forces to round up and brand livestock.[5] Similarly, miners in the early West set up mining districts, formed associations, and established courts in order to acquire and retain their mining claims and to appropriate water needed to work them.[6] Indeed, development of these institutions hinged on cooperation.

Yet, just two approaches have dominated the way the tragedy of the commons is avoided—regulation and property rights. Both are grounded on the argument that cooperation among users is not possible. The first calls for turning management of the commons over to the government, with its "major coercive powers."[7] Unfortunately, as we have seen for public lands, fisheries, and pollution, the regulatory approach has a tendency to promote wasteful production and often is equally destructive to the environment. The second calls for privatizing the commons. Not surprisingly, this approach has been shown to lead to lower production costs, much higher incomes, and greater conservation for relatively stationary resources, such as oysters.[8] However, individual ownership is not feasible in every situation. It may be too costly for some resources, or it may be socially unacceptable.[9]

Often omitted from policy prescriptions is the possibility that a nongovernmental community of users can limit entry and control resource use. In many places, local users manage common areas, usually without governmental interference, and they prevent overexploitation. In many cases, these arrangements are "community-based, spontaneously developed and informally organized."[10]

Such an arrangement is a "practical option if a resource is sufficiently valuable to justify the costs of organizing the group but not the costs of defining, establishing, enforcing, and exchanging private rights."[11] It has worked well throughout the world in managing many resources, including mountain forests and meadows, water for irrigation, coastal fisheries, and more recently, recreation and wildlife. The very existence of these community-managed commons challenges the assumption that users are locked into a destructive pattern of competition that invariably leads to resource abuse. Community management shows how the commons can be self-regulated. Scholarly attention to such self-regulated commons has grown in recent years, providing us with valuable information on why such management can occur.

CUSTOMS, CULTURE, AND THE COMMONS

Until recently, traditional economic analysis has given scant attention to the potential for societal norms as constraints on human behavior, on the grounds that human nature is inherently self-interested.[12] But examples to the contrary have forced institutional economists to confront the prospects of customs and culture as constraints on human behavior. In the absence of either governmental control or individual ownership, not all resources held in common are subject to the tragedy of the commons.[13] In the Swiss alpine regions, for example, communally owned lands have remained productive for grazing and logging for centuries. The rules governing access and use of these lands are devised by villagers.[14] Similarly, in Japan, local village institutions have regulated grazing and logging on commonly held meadows and forests for centuries. None of these has "suffered ecological destruction."[15] For at least five centuries, irrigators near Valencia, Spain, have relied on local rules to allocate water, local arenas for resolving water disputes, and fines to sanction violators. In this semiarid region of highly variable rainfall, the system has proven more than adequate for resolving water disputes and allocating water reliably in both dry and wet years.[16]

In light of such examples, political economists have become increasingly interested in customs and culture as constraints on human action. In her analysis of the commons, Elinor Ostrom considers the constraints of societal norms that she believes

> reflect valuations that individuals place on actions or strategies in and of themselves, not as they are connected to immediate consequences. When an individual has strongly internalized a norm related to keeping promises, for example, the individual suffers shame and guilt when a personal promise is broken. If the norm is shared with others, the individual is also subject to considerable social censure for taking an action considered to be wrong by others.[17]

Though the exact conditions under which societal norms arise are not yet well understood, several factors characterize community management systems that have survived anywhere from one hundred to a thousand years overcoming various social, political, and economic changes.[18] As such, they are considered desirable design features for community management. A list of these factors follows.

1. *Boundaries must be clearly defined so that individuals within a group know which resources they can use and how, and so that individuals outside the group know when they are trespassing.* This factor is viewed as a critical step in promoting collective action on the part of users sharing the commons. So long as boundaries are uncertain, no one knows what is being managed and for whom. Moreover, local users face the threat that any benefits achieved through community management run the risk of being captured by outsiders. The presence of clearly defined boundaries is what distinguishes this form of common property from one that is subject to open access.
2. *Group decisions require rules that determine how the group parcels out the value of the resource.* Simply defining the boundaries of a resource is not enough, however, because it is still possible for a limited number of appropriators to increase extraction rates so that they dissipate potential rents or totally destroy the resource itself. To avoid these outcomes, communities must develop rules of appropriation, and users must adhere to these rules.
3. *Customary rules must be linked to the time-and-place-specific resource constraints so that resulting rules are efficient, or there will be pressure to change them.* Tailoring rules to local conditions is another important contributor to robustness and longevity. Such rules take into account specific attributes of the resource being exploited, local economic and political conditions, and cultural views.
4. *There must be effective monitoring of rules, and rewards for individuals who abide by rules, or sanctions against those who violate them.* Ostrom notes that "even in repeated settings where reputation is important and where individuals share the norm of keeping agreements, reputations and shared norms are insufficient by themselves to produce stable cooperative behavior over the long run."[19] She goes on to note that there is indeed "substantial evidence" that participants in long-enduring community management systems carry out effective monitoring and sanctioning activities to ensure rule compliance.[20] Moreover, when appropriators design and enforce their own rules, they learn from personal experience which rules work and which do not, and which rules achieve the highest net benefits.

5. *Where conflicting demands are likely to arise between group members, dispute-resolution mechanisms such as local arenas for bargaining are necessary.*[21] Even in smooth-functioning groups, disagreements are inevitable. A mechanism for settling such disagreements is critical to maintaining cooperation on the commons. In community systems where the likelihood of conflict is high, well-developed court mechanisms have been in place for centuries.
6. *The rules must not be subject to change by higher levels of government.* According to Ostrom, when community management systems devise rules that lack legal authority to back them, they remain fragile at best.[22] The danger is that outsiders may use the government to overturn the rules devised by the community.

For these six factors to emerge and to be maintained on the community commons, members must perceive themselves as having strong group identity. Group homogeneity will help promote this identity. For such a group, "outcomes to other group members, or to the group as a whole, come to be perceived as one's own."[23] Group identity reduces the efforts that people must devote to centralized authority or to definition and enforcement of private property rights, but it requires that members invest in maintaining societal norms and in keeping out people who do not share these norms.

In the modern context, the controversy over restricting immigration illustrates how costly this can be. If group identity is to be preserved, resources must be invested in limiting entry to individuals willing to accept the group's standards of conduct and to produce and maintain customs and culture. Tests for initial acceptance include such mechanisms as knowledge of the group's history, language requirements, and residency requirements. Rituals, ceremonies, and formal and informal education also provide ways of inculcating customs and culture.

Ultimately, however, limiting entry requires the ability to threaten credible force to exclude individuals not sharing the group's perspective. This requires collective action by the members that may be as simple as excluding an individual from some collective activity with significant economies of scale or as complex as mustering arms and threatening war. The modern nation-state is the main mechanism for the latter role, but informal groups such as producer associations and social clubs play a role in excluding nonmembers and enforcing societal norms.[24] However, these informal organizations may lack the legal authority to exclude would-be entrants.

Outsiders wishing to enter the commons have three choices: they can negotiate with the group for permission to use the resource, they can force their way in, or they can trade.[25] All three make customary constraints tenuous at best. In the case of trade, once property rights change hands, the new owner may not share the group's perspective. Preventing trade with outsiders or requiring strict standards of acceptance for entry are two methods communities use to ensure integrity of the group. Such restrictions may be discomforting to some in the sense that those who value the resource may not be able to trade freely for it, but such restrictions must be weighed against the net benefits generated by the community in making local rules work.

LESSONS FROM AMERICAN INDIANS

The coastal Tlingit and Haida Indians of pre-white Alaska had strong incentives to avoid the tragedy of the commons for sockeye salmon. Sockeyes differ from chinook, chum, and other salmon in that they arrive early in spawning streams, remain there the longest, and show the lowest variation in numbers returning from one year to the next. Duration and stability, coupled with high nutritional value, explain why sockeyes were an important food source for the Indians. Their value as a food source encouraged Indians to establish clear boundaries specifying who had access to the stream systems where sockeyes congregated on their journey to spawning beds.[26] Access to these locations was limited to the clan or house group, the social organization that could be supported by the resource. The optimal size of organization that possessed customary fishing rights to the mouth of a stream was determined mainly by the number of individuals needed to deploy most traps and weirs. This typically entailed five to eight men, the size of the house group. However, very small creeks could be "the special preserves of individuals."[27]

In addition to property rights "to the all-important salmon streams," resource territories included "bear- and goat-hunting areas, berry and root patches, hot springs, sea otter grounds, seal and sea lion rocks, shellfish beds, cedar stands and trade routes."[28] These units could exclude other clans or houses from their territories, thus localizing management decisions and restricting the potential for capturing the value of the resource to members of the clan or house. When territories were infringed upon, the trespasser was required to indemnify the owning group or potentially face violent consequences.[29]

The totem pole and the potlatch system were two legendary institutional mechanisms that served to substantiate a claim of an individual clan head to a salmon stream or territory.[30] "Totems of the potlatch . . . buttressed the resource management system of the Tlingit by symbolically representing clan accounts and claims to salmon streams."[31] Potlatching also appears to have strengthened enforcement, because it involved competitive reciprocal gift giving among clan leaders and ensured that leaders whose salmon streams or territory were temporarily unproductive would have more to gain by respecting their rivals' property rights than by encroaching.[32]

Northwest coastal Indians did not define and enforce property rights to resources that were not scarce, which is consistent with property rights theory.[33] The Tlingit did not establish territorial claims to streams where species such as the pink or dog salmon were abundant, and they also treated the open ocean as a commons because their technology did not allow overexploitation when salmon were in this environment. Some bands, however, did claim ownership of bottom fishing grounds for two highly valued food sources, halibut and cod.

Management decisions were linked to time-and-place-specific resource constraints through the eldest clan male, or the *yitsati*. This person generally possessed superior knowledge about salmon runs, escapement, and fishing technology and therefore was in the best position to be the "custodian or trustee of the hunting and fishing territories."[34] The *yitsati,* or "keeper of the house," had the power to make and enforce rules regarding harvest levels, escapement, fishing seasons, and harvest methods. Though there is debate over how powerful the *yitsati* was, it is clear that salmon runs were sustained over long periods by local rules that parceled out the rents from the resource, and that these rules took into account time-and-place-specific knowledge. The *yitsati* "had the power of life and death in enforcing these regulations," thus guaranteeing that self-interest not constrained by societal norms could be sanctioned by collective force.[35] The *yitsati* also assisted in parceling out goods produced collectively to members of the clan.[36]

To minimize enforcement and transaction costs within the proprietary groups, rights initially could not be transferred to those outside the clan.[37] This allowed those with proprietary interest in a fishing location to exclude people who might not abide by customary norms, thus limiting self-interest.

Such limits on transfer conflicted with European notions of property in a heterogeneous society and restricted Europeans' access to the resource. The Russians were the first to confront Indian fishing rights. Acting as a monopolist, the Russian American Company denied individuals the right to fish, thereby making itself the sole negotiator with the Indians. The company then had the choice of using force to take fishing rights from the Indians or of trading with them. It used both, depending on relative military strengths, transaction costs, and the value of the salmon fishery in question.[38]

When the United States purchased Alaska from the Russians in 1867, leasing gave way to military conquest. "The major purpose of the military in retrospect was to protect U.S. citizens in their encroachment on Tlingit resources and impose U.S. legal concepts. This period has been termed the era of flagrant neglect in Alaskan history."[39] Indeed, by neglecting the authority of Tlingit to establish fishing rights and substituting the public trust doctrine from English common law, the U.S. government created the tragedy of the commons. "[B]y 1900 American common property principles were in force throughout southeast Alaska, at least insofar as the Tlingit and Haida were unable to keep Euro-Americans from fishing where they wanted to commercially."[40] By allowing anyone to place traps and weirs at the mouths of rivers, there was little incentive to worry about escapement; a fish left to spawn was a fish potentially caught by someone else.

Not surprisingly, the result of ignoring native fishing rights was the same "legally induced technical regress" described by Robert Higgs for the Washington fishery.[41] As noted in chapter 9, the Indians along the Columbia had well-established rights to fishing sites and had developed effective technologies for harvesting salmon. These rights were usurped when state and federal governments allowed newcomers to place their nets across the Columbia, ultimately decimating salmon populations and leading to state regulation.

Instead of acknowledging the customary native fishing rights, the legal system encouraged a race to catch fish where fish gathered to spawn. Because this led to quick depletion of salmon, traps and weirs were banned only to be replaced by purse seine boats powered by internal combustion engines. Fishing for salmon moved to open waters. Ironically, from the country where property rights were considered a fundamental part of law came a legal system driven by regulation.

CONTEMPORARY COMMUNITY FISHERIES

In recent years, researchers have uncovered a host of contemporary examples of community-managed fisheries. For example, in the coastal waters off Alanya, Turkey, unconstrained competition among local fishers for the best fishing spots led to conflict and, at times, violence. In addition, competition for the choicest fishing spots increased production costs and harvest uncertainty for individual fishers.[42]

In the early 1970s, a local producers' cooperative whose members included half the fishers in the fishery began experimenting with a set of rules for assigning fishing spots to all eligible fishers, consisting of all of Alanya's licensed fishers, regardless of membership in the co-op. Ten years of refining the set of rules resulted in a highly effective system that minimizes the incentive to compete for the choicest fishing spots.

Before each season, a list of all eligible fishers is prepared, along with a list of fishing locations based on the most recent information from fishers. Fishers then draw lots for their initial assignment on opening day of the fishing season. Every day thereafter, each fisher moves east to the next site until the end of January. After January, each fisher reverses course and moves west to the next site. This gives everyone about the same opportunity to reach the stocks of fish, which migrate from east to west between September and January and from west to east from January to May. This arrangement eliminated the need to fight over prime fishing sites and reduced the incentive for overcapitalization—that is, excessive investment in fishing gear to capture fish before anyone else does.[43]

Although fishing sites are not individually owned, rights to use fishing sites are well defined and enforced by the fishers themselves. And although the Turkish government does not regulate the fishery, national legislation that has given such cooperatives jurisdiction over local arrangements legitimizes the role of the cooperative in devising a workable set of rules for fishing.

Still, it is worth noting that Alanya's innovative system for assigning fishing sites does not address the problem of limiting access to the fishery. For that reason, as a system capable of withstanding the test of time, it must be considered fragile. At the current time, the number of eligible fishers—all licensed fishers in Alanya—does not lead to overfishing. But if more individuals decide to obtain licenses and enter the fishery, overfishing could well occur.

The Mawelle fishery, another Turkish fishery with local fishing rules and monitoring, succumbed to overfishing because it failed to limit access. Ostrom identifies a number of community systems that either have unraveled and allowed overexploitation to occur or are at high risk of unraveling. In each case, a few of the factors that characterize long-lasting community systems (outlined earlier) are missing.[44]

The large network of Fishing Cooperative Associations (FCAs) governing much of Japan's nearshore fisheries offers another example of a government-recognized system. By law, FCAs own the fishing rights to specific territories extending as much as five and a half miles seaward.[45] About 5,000 associations are scattered around Japan's coast. These associations manage the fishery resources subject to guidelines and conditions set down by national and regional governments. To fish a specific area under jurisdiction of a cooperative, a fisher must be a member and must comply with its rules or risk being expelled from the organization and its fishing area. The overall benefits of these organizations can be seen in "the stability of coastal catches" and fisher incomes that are "equal to or above the national average" for all workers in Japan.[46]

Historically, these cooperatives have their roots in community customary law and the formal laws of Japan's feudal era—from 1603 to 1867. By feudal times, the increasing number of fishers in the coastal areas led to many disputes within and among coastal villages. To help resolve these disputes, the feudal lords granted territorial fishery rights to village guilds and encouraged the guilds to work out solutions among themselves. Further refinements in the system have taken place since then. In 1884, the government enacted "Working Rules for Fishermen's Associations," which transferred fishery management from the village guild to a "fishery association" made up of local heads of fishing families.[47] In 1941, the fishery association became known as the fishing cooperative. Sweeping reforms shortly after World War II enabled all fishers from a local community to be eligible for cooperative membership, not just heads of families.

Although the rules have legal backing, adherence to them within a cooperative still depends very much on the cohesiveness of the local community. Kenneth Ruddle and Tomoya Akimichi conclude that "community norms are flouted at one's peril and threat of social banishment (*murachibu*) is real and horrifying."[48] On the other hand, the anonymous regulations established by the prefecture (regional government)—and even more so those imposed by the national government—are, in general, perceived as being less binding. Kevin Short emphasizes the importance of community cohesiveness:

> Befu (1980)[49] has documented examples of cutthroat competition and frequent conflict among fishermen of the Inland Sea region. What, then, prevents similar conditions from occurring in Shukutsu? The answer, I believe, lies in the social and cultural forces that bind the fishermen into a relatively tight-knit, cohesive group in which individuals are willing to compromise their interests for the sake of their group.[50]

Many community fisheries lack the government recognition of the previous two examples. Maine's Matinicus Island community lobster fishery has operated successfully for over a century without official state recognition, despite many changes—including expansion into regional markets and dramatic improvements in boats and fishing technology.[51] And while the number of fishers has deviated little from the original number of 36, fishers move in and out of the fishery.[52]

Island fishers strictly control who will be accepted into their fishery. One must either live on the island and have island kinship or purchase property from a local fisher, who then becomes an informal sponsor. The latter approach is akin to an apprenticeship program. In addition, one must demonstrate a willingness to cooperate with other fishers and respect their fishing rights and equipment. An individual must also make the necessary investment in wharf access, boat, and traps, an investment that totaled roughly $125,000 in the 1980s.[53]

Local fishers actively defend the territory from outsiders through extralegal means. In their study of the fishery, anthropologists Frances P. and Margaret C. Bowles write:

> They [local fishers] customarily signal a territory violation by opening the door and tying a half-hitch around the buoy of an outsider's trap. If this signal is ignored, an island lobsterman may haul up the outsider's trap and dump them together so that the buoys and warps become tangled. Actual trap cutting ensues only if these measures fail to convey the wisdom of removing the offending gear from the disputed area.[54]

On well-defended waters like those off Matinicus, lobster fishers have instituted their own conservation efforts, which have benefited both lobsters and fishers. "On Matinicus and Green Island fishermen have agreed to limit the number of traps used,"[55] notes anthropologist James Acheson. In comparison with more open lobstering areas off Maine, Acheson found that well-defended areas such as the waters off Matinicus have substantially higher lobster density and fisher incomes.[56]

Community rules for lobstering in the waters off Matinicus have worked because island fishers have been able to restrict entry and introduce their own conservation measures for over a century. Still, Acheson notes that the state government of Maine could "annihilate the entire territorial system if it so chose by vigorously enforcing laws concerning trap cutting."[57] Community management exists, says Acheson, "only because of the benign neglect of the state."[58]

RECREATIONAL SOLUTIONS

Communities can protect recreational fisheries as well as commercial ones. For example, the provincial government of Quebec tried for years to limit salmon gillnetting by the Micmac Indians at the mouth of the Cascapedia River, to no avail. Then, in 1992, it took a different tack. The government allowed a local board to take over management of the river's salmon fishery, with financing from Cascapedia user fees. Notably, half the board members were Micmacs, and half were local sportsmen and other community members. Over the next two years, a spirit of cooperation took hold. The Micmacs greatly reduced their gillnetting of salmon and, in return, were trained as river guides and river guardians. Today more than one hundred Micmacs are employed as guides and private wardens and in other positions that support the salmon sportfishery on the Cascapedia River. The income from these services has more than compensated for the reduced gillnetting. Salmon caught commercially are worth about $30, but salmon caught by sportfishers are worth $400.[59] Harmony, conservation, and lucrative sportfishing are the by-products of this locally managed fishery.

The Cascapedia story is not the only community solution to recreation in Quebec. Quebec also has ZECs (Zones d'Exploitation Contrôlées) for hunting and fishing. A ZEC is an area of Crown land (land owned by the province) or section of a salmon river in which a locally based nonprofit corporation receives authority from the provincial government to develop and control recreation, set and collect fees for access and hunting and fishing, monitor fish and wildlife populations, and set and enforce seasons and bag limits in conjunction with government guidelines. The goal of each ZEC corporation is to become financially self-sufficient.

In 1978, the first year of the ZEC program, the Ministère de l'Environnement et de la Fauna (Ministry of Environment and Wildlife) created 55 hunting and fishing ZECs, which included hunting for such species as bear, moose, and deer and fishing for such species as walleye and trout. A few years later, the government established several salmon ZECs to capitalize on the growing demand for salmon sportfishing and more recently a waterfowl ZEC. As of 1995, there were 82 ZECs, consisting of 63 hunting and fishing ZECs, 18 salmon ZECs, and one waterfowl ZEC.[60]

According to Yannick Routhier of the Ministère de l'Environnement et de la Fauna, the ZEC is a kind of co-op. ZEC users, through elected representatives who serve on the managing board of directors of the corporations, have a voice in managing recreational use and controlling wildlife resources.[61] Many ZEC boards have broadened their composition to include local business interests and tribal interests. For example, the Rivere des Ecoumins, a recreational salmon fishery located 150 miles northeast of Quebec City, has a managing board of which one-third are recreational users, one-third local municipalities, and one-third Montagnais Indians.

In comparison with national forests in the United States, where hunting and fishing and road use are priced at zero, fees are charged for ZEC membership ($15) and road use ($4 per day and $30 per year). Fees for hunting and fishing vary. The average daily fishing fee for species other than Atlantic salmon ranges from $8.75 to $11. The daily rate for fishing in a salmon ZEC, the *crème de la crème* of sportfishing, runs as high as $55. Fees also vary with exclusivity of use. In Ste. Marguerite salmon ZEC, for example, a ZEC member pays a daily salmon-fishing fee of $23 on sections where the number of rods is unlimited, but nearly twice that amount on sections where the number of rods is limited. Nonmembers enjoy the same access privileges but at slightly higher user fees. A number of ZECs have been able to completely offset operating costs out of fees. For example, gross income from fees on the Ste. Marguerite was about $100,000 in 1994, which about covered expenses for maintaining roads, providing private wardens, monitoring wildlife numbers, and carrying out conservation projects.[62]

In addition to financial progress, ZECs are helping improve resource management and maintain recreational quality. Jean-Francois Davignon of the Atlantic Salmon Federation notes widespread improvement in protecting, monitoring, and conserving salmon fisheries due to ZEC local support. He points out that ZEC managing corporations hire 116 wardens from local communities to assist the government in protecting wildlife and fisheries. For salmon rivers on the Gaspe Peninsula, salmon-spawning runs have shown marked improvement under ZEC management. "In 1984," says Davignon, "Gaspe peninsula rivers were only at 30 percent of required salmon spawners. Today, they are averaging between 80 percent and 100 percent."[63] In addition, as the ZEC system developed and pressure on resources rose, ZEC corporations have rationed use by limiting the number of hunters and fishers in certain areas and raising fees. Such efforts have been instrumental in maintaining high-quality recreation in the system.[64]

RESOLVING THE WILDLIFE COMMONS

In recent years, community systems have become a powerful tool for solving wildlife problems in impoverished areas of the world.[65] Consider the situation in southern Africa. Idyllic scenes of wild animals freely roaming vast savannahs are disappearing because often those animals compete with humans struggling to survive on the same turf. Although some southern African nations have set aside 18 percent of their land as national parks and wildlife preserves, human population pressure is reducing the wildlife habitat, and poaching is depleting wildlife herds.[66]

Wildlife problems result because indigenous people trying to survive off the land must compete with wildlife; hence, wildlife is a liability rather than an asset. Understandably, these people resent the fact that land is set aside for wildlife instead of them. Moreover, out of the preserves and parks come lions, leopards, and elephants that venture onto lands used for human subsistence, where they destroy crops and livestock and occasionally people. Add to this the problem of insufficient public funds for environmental protection and wildlife problems become practically impossible to resolve through the traditional Western approaches, such as public maintenance of habitats and governmental control of wildlife.

Entrepreneurs who understand the problems of wildlife in southern Africa are working with local communities and national governments to change the incentives indigenous people face when it comes to wildlife. Sixty to 80 percent of Africa's people live in rural areas and most of them barely get by on subsistence farming and ranching on communal lands.[67] Many of these lands are marginal for agricultural use but provide excellent wildlife habitat. The problem is that sustainable wildlife has not meant sustainable communities.

Zimbabwe is trying to change this through the innovative Communal Areas Management Programme for Indigenous Resources (CAMPFIRE). CAMPFIRE seeks to conserve wildlife by creating positive incentives for human communities to coexist with wild animals. These incentives entail capturing returns from wildlife enterprises, such as hunting and photographic safaris; expanded job opportunities for locals in resource management; and greater utilization of wildlife by-products such as meat and hides for local purposes. Subject to government guidelines for maximum allowable harvests, CAMPFIRE effectively devolves the rights to use, manage, and benefit from local wildlife to district councils and the communities they represent.[68] Since these rights are exclusive, district councils and their communities are free to capture wildlife values now or in the future without worry that what is not captured from wildlife today will be taken by others tomorrow (thus reducing the potential for the tragedy of the commons). In other words, these entities are better able to capture returns from wildlife, losing revenue only to the extent that they allow poaching or fail to provide productive habitat conditions.

As an example of how a CAMPFIRE project can work, the Nyaminyami District receives funds and technical support from a local nonprofit development organization to help assess local wildlife conditions and develop a plan of utilization. With this support, the Nyaminyami District Council submits a local wildlife count to the national wildlife department each year. Upon verification, the district council is given a maximum allowable quota equal to one percent of the local wildlife count to hunt or cull wildlife for local purposes. In 1989, its inaugural year, this CAMPFIRE project earned enough from safari hunting and culling efforts to cover the project's operating and administrative costs and provide $6,400[69] to local communities.[70] This amount might not seem like much by Western standards, but realize that the average annual cash income per household in the district was less than $100 in 1989.

The Beitbridge District provides another illustration of early CAMPFIRE success. Unlike Nyaminyami's project, however, the Beitbridge District CAMPFIRE project was financed entirely out of wildlife returns. In 1990, the Beitbridge project generated approximately U.S.$20,000 from safari hunting sales. Notably, the Beitbridge District Council chose to distribute the wildlife returns to communities based on the amount of wildlife produced in their respective communal land areas. It allocated 87 percent of the total returns to the community of Chikwarakwara, which was the top wildlife producer among all the communities in the district. National parks also paid the Beitbridge District Council $18,400 for proceeds accrued from past safari hunting, and from this amount, Chikwarakwara received an additional $8,000. Free to choose how to use the proceeds, the people of Chikwarakwara decided to pay each of the 149 households in the community $80 as a wildlife dividend. The remainder of the earnings went toward funding of a school and purchasing a corn-grinding mill. Again, to wealthy Westerners, the $80 dividend may not seem like much, but it almost doubled the average annual cash income of each family. "Now the people of Beitbridge are reported to be talking seriously about how to control poaching. They are considering the possibility of reducing a household's cash payment by the value of any animal poached by any of its members, e.g., Z$75 (U.S.$30) for an impala."[71]

Other districts in Zimbabwe have followed these early CAMPFIRE successes with projects of their own. The Binga District project entails a lease with a private hunting safari operator and joint ventures with two photographic safari operators. In addition, plans are under way for a commercial fishing venture at the western end of Lake Kariba. The Hwange District is developing the "scenic attractions and natural resources" near the Zambezi River and Victoria Falls by forming joint ventures with two photographic safari operators.[72] The Guruve District includes an arrangement in which crocodile farmers pay communities along the Manyame River a fee for all the eggs collected, thereby stimulating community protection of crocodile nests. As of 1995, 25 communal land districts adopted CAMPFIRE projects, and 13 of them generate significant revenues from wildlife.[73]

How effective these projects are in promoting wildlife conservation depends on whether local communities bearing the brunt of wildlife costs are allowed to benefit from their management efforts. One problem that appears to have been addressed recently was the CITES[74] ban on the ivory trade.[75] This policy had all but eliminated one lucrative source of revenue for CAMPFIRE communities. In addition, some evidence suggests that bureaucrats in the central government are beginning to siphon off some of the returns from wildlife. If this continues, the incentives for community stewardship would be severely weakened.

GUARDING FORESTS

In addition to wildlife, communal arrangements have also proven effective in protecting forest values. For example, in Torbel, a village in the Swiss Alps, villagers have held a combination of private and common property rights to natural resources for centuries. On lands that are productive for growing grain, hay, vegetables, fruit trees, and other crops, property rights are privately held. In contrast, on forests located in areas characterized by steep, rugged terrain, low precipitation, and short growing seasons, timber returns are too low to justify private ownership. Instead, these forests are held in common by villagers for wood and other uses. Even though these forests are not highly productive for timber, they protect the village from avalanches and help maintain an aquifer. As a result, villagers assess forest conditions and impose a set of rules to ensure their forests are not overharvested.[76]

Similarly, in New Guinea a communal land rights system held by local tribes has helped protect rain forest amenities, including myriad rare bird species. Here one tribal clan or another holds property rights to every inch of rain forest, "no matter how remote or how worthless."[77] These rights have been passed down from generation to generation and cannot be transferred permanently to outsiders. Tribes defend their forest territories against trespass by outsiders and control how forest and wildlife resources are used with local rules.[78]

More recently, this form of community ownership has enhanced rain forest protection in the midst of oil development. When oil was discovered in the Kikori River area in the late 1980s, various tribes exercised their communal rights and demanded compensation. Jared Diamond describes the local response when a subsidiary of Chevron made a clearing on the top of a mountain peak as a helipad:

> . . . as soon as Huli tribesmen living around the base of the mountain saw helicopters hovering there, they began climbing. On reaching the helipad,

> they introduced themselves as representatives of various clans, and all announced their claim to the previously worthless spot, demanding huge compensation for trespassing, cutting trees, and disturbing birds. Local Samberigi tribesmen went further by later hijacking chartered helicopters to enforce their demands. Faced with a hornet's nest of lawsuits and bills for damages, Chevron abandoned the place.[79]

Because locals held strong communal rights to rain forest areas, there was a concerted effort to ensure environmentally sensitive oil development. Notably, at the time of oil development there were no relevant government regulations for protecting New Guinea's rain forests. Yet the oil developer devised a comprehensive environmental plan to ensure that there would be minimal disruption to the delicate rain forest. The plan included an assessment of the socioeconomic, cultural, and archeological impacts of oil production, as well as a number of precautionary measures. For example, the base camp, airfield, and access roads were all located outside the drainage area of the ecologically sensitive Lake Kutibu. Project supplies were delivered by boat or airplane to reduce the construction of roads. And much of the vegetation disrupted by the construction and drilling has been either replanted or allowed to grow back on its own. Following the start of oil production, World Wildlife Fund scientists studied the biodiversity of the area and found no adverse effects from the oil project. Local tribes not only get royalties on all oil pumped from their lands, they also get a significant amount of goodwill from the developer. Chevron has built schools, hospitals, and roads for their use.[80]

In contrast to Torbel's and New Guinea's forests, Nepal's forest experienced several decades of destruction because the national government chose to ignore community arrangements. In 1957, the government placed all the nation's forests under central control of the government. The act led to a "chain of destruction" that resulted in the removal of almost half of the trees in Nepal's forests.[81] The nationalization of forests prompted villagers to harvest forests at an unsustainable rate.

> Nepalese villagers began free riding—systematically overexploiting their forest resources on a large scale. The usual explanations for this free riding are that the villagers had lost control of their forests, and they were distrustful of government control and national resources policy.[82]

In recent years Nepal has been trying to turn the tide in its diminishing forests. An amendment to the National Parks and Wildlife Conservation Act of 1973 authorized park officials to create local user committees to participate in managing park resources, while providing that between 30 to 50 percent of park revenues would be distributed to local communities. In addition, the government handed over 216,000 hectares of forests to local communities.[83] These communities are allowed to harvest wood and animal feed from their forests for their own subsistence. When the forests mature, they may also selectively cut trees to sell for profit so long as they plant ten new trees for each tree harvested.[84]

Unfortunately, Nepal's community forest program has several drawbacks that limit its effectiveness. First, local communities do not enjoy complete decision-making authority over use of their forests. The government's park warden has the power to dissolve local forest committees that do not perform efficiently, deny permission to cut trees or extract other resources, and issue or deny permits for all commercial activities on forestlands.[85] In addition, there is a problem with dispersal of funds earned from nearby parks. These funds are designed to induce locals to protect forest resources linked to the health of nearby park resources. By protecting forest resources for the benefit of nearby park resources, locals stand to benefit financially. However, there is no requirement that park funds be returned to an area where they were earned.[86] The government could disburse funds to communities in an entirely different region. As a result, Nepal's program for community-managed forests is one that lacks the same incentives for protection generated under New Guinea's strong regime of communal land rights.

CONCLUSION

While the tragedy of the commons often results when there are many users who differ widely in assets, such as harvest skills, knowledge, ethnicity, or other group-dividing variables, there are cases where a group of users, sharing strong group identity, prevents the tragedy of the commons. They present fertile ground for research by institutional economists and policy analysts if for no other reason than that it challenges the notion that without centralized governmental control or pure privatization of resources, users are locked into a destructive pattern of use that results in the tragedy of the commons.

Community systems have a long history of use in coastal fisheries, mountain forests and pastures, and irrigation systems. More recently, they have been applied to wildlife and outdoor recreation. It remains to be seen whether these new applications, which share decision-making power and benefits with government, can continue to exhibit success.

Whatever the application, there is strong evidence that a combination of six factors contribute to system longevity. All of the systems that have lasted centuries have exhibited all the six factors outlined at the beginning of this chapter. In contrast, community systems with some of these factors missing are either at risk of unraveling or have already failed to prevent the tragedy of the commons.

CHAPTER 12

TAKING FREE MARKET ENVIRONMENTALISM GLOBAL

Most of this book has focused on U.S. environmental policy, but increasingly the environmental movement has taken environmental issues global.[1] Domestic policy regarding carbon emissions, foreign policy regarding military operations, international trade agreements such as the North American Free Trade Agreement (NAFTA) and the General Agreement on Trade and Tariffs (GATT), United Nations conferences and policies, and international law have all taken on an environmental flavor. The protests surrounding the World Trade Organization meetings in Seattle in December 1999 and the meetings of the International Monetary Fund and the World Bank in April 2000 illustrate the intensity of feelings about the interface between international relations and environmental quality.

Like domestic environmental policy, the global environmental movement has focused on governmental regulations. To mention a few, the Montreal Protocol regulates chlorofluorocarbons (CFCs) by stipulating when countries will phase out these potentially ozone-damaging compounds. The Kyoto Protocol regulates greenhouse gases by stipulating when countries will have to cut emissions of these gases to roughly 95 percent of 1990 levels. This will happen mostly by reducing the use of fossil fuels that emit large quantities of carbon dioxide into the atmosphere. And the Convention on International Trade in Endangered Species (CITES) specifies which animal, fish, and plant species are in danger of extinction and strictly regulates whether they can be traded across international borders.

Though regulations are in place or have been proposed for numerous global environmental issues, there is significant scientific debate over the extent and consequences of the problems in the first place. At the pinnacle of such debates is global warming, with scientists lined up on both sides armed with models and data. This debate began with atmospheric scientists using computer models that predicted dramatic increases in global temperatures due to a steady buildup in carbon dioxide and other trace greenhouse gases in the atmosphere. The theory behind the models is not without challenge from experts who argue that scientific proof is incomplete or contradictory, and that there remain uncertainties about the nature and direction of the Earth's climate.[2] Also, possible impacts from global warming are seen by some experts as negative and by others as positive.[3] In any case, if it materializes, a warmer climate would probably have far-reaching effects on crops and forests, weather patterns, marine life, human health, water resources, and sea level. Since the initial predictions, debate has refined the models and the data used to test the predictions from the models, and as a result, the magnitude of global warming by 2100 has been ratcheted downward.[4]

Because this book is about policy, however, we will leave the science debate to scientists, and suffice it to say that such debates are crucial to good policy. Before leaping into any policy prescriptions, we should know whether the cure is more costly than the disease, if indeed the disease exists. Just as defining and enforcing property rights to the grazing commons on the western frontier did not make sense until the effects of crowding began to manifest themselves (see chapter 3), regulating the global commons can generate costs in excess of benefits. In the case of global warming, for example, with the jury still out on the impact of greenhouse gas accumulations, the best estimates at this time suggest that temperatures may rise by about one degree Celsius by 2050. No one questions whether doing what is prescribed in the Kyoto agreement will have substantial costs. Total emissions of carbon dioxide and other greenhouse gases (mainly methane and nitrous oxide) are much greater that they were in 1990. As a result, industrialized nations, and especially the United States, face a daunting and perhaps impossible task. To comply with Kyoto regulations, the United States must reduce its emissions projected for the deadline period by some 40 percent.[5] But will it be worth incurring these costs? The best guess is that doing everything proposed in the Kyoto accord will reduce the predicted rise in temperature by only one-tenth of one degree by 2050 and that such a reduction will hardly reverse the consequences of global warming.[6]

The important point is that good policy cannot lose sight of the additional benefits and costs of taking action. Or, putting a small twist on an old adage, anything not worth doing is certainly not worth doing well. But when the science does suggest that a tragedy of the global commons may be imminent, how can free market environmentalism be taken global? The answer to this question requires sticking to the basics of the paradigm, namely the importance of property rights and the rule of law.

IS SUSTAINABLE DEVELOPMENT THE ANSWER?

Perhaps the most market-oriented policy approach coming from the globalization of environmental issues is what is called sustainable development. According to this prescription, "the current generation must not compromise the ability of future generations to meet their 'material needs' and to enjoy a healthy environment."[7] Although there are different interpretations of sustainability, advocates generally

- perceive that the biosphere imposes limits on economic growth,
- express a lack of faith in either science or technology to lead to human betterment,
- are extremely averse to environmental risks,
- support redistributive justice and egalitarian ethics,
- profess concern over population growth and have faith in the wisdom of human capital development (education), and
- have survival of species and protection of the environment and of minority cultures, rather than economic growth per se, as goals.[8]

Sustainable development, as advocated by today's ecological economists, is a holdover from the 1960s and 1970s, when economists were struggling with steady-state and zero-growth economic models. These models used the intuitive concept of the "economics of the coming spaceship Earth" to get people thinking about what was necessary to sustain life on earth. With strict limits on food, air, and energy in the spaceship, rationing is unavoidable and that rationing requires regulations, not markets. By this analogy, achieving what used to be called a "steady-state economy,"[9] now sustainable development, requires strict political controls that carefully balance product consumption, energy use, and wastes. During the 1970s, the demands for strict political control of resource consumption were driven by a concern that energy resources were being exhausted. Although all indications tell us that price deregulation solved the energy crisis, the steady-state theories formed the basis for many regulations, from climate control in buildings to lower speed limits to fuel efficiency standards and ultimately to a new bureaucracy, the Department of Energy. Today, the same regulatory zeal under the guise of sustainable development is being driven by fears of global warming and biodiversity loss.

The politics of regulation necessary to implement sustainable-development policies are likely to dwarf problems that are inherent in national environmental regulations. At the national level, there are governmental institutions that can implement regulations, and in democracies there are checks and balances that control regulators. Global regulation, however, would require international treaties or organizations to specify and implement the regulations. But this raises important and complex questions of who will come to the bargaining table and who will negotiate with whom. If agreement is reached, who will do the enforcing? If fines or taxes are to be imposed, who will collect them and who will receive the proceeds? These are serious questions that have not been adequately addressed by the advocates of global regulation.[10]

Unfortunately, as Timothy O'Riordan points out, sustainable development's "beguiling simplicity and apparently self-evident meaning have obscured its inherent ambiguity."[11] Hence the concept has become everything to all people. One version of sustainability (which is not inconsistent with neoclassical economics) calls for "maximizing subject to constraints." According to this version, all ecological principles and environmental ethics must be taken into account by the institutional framework that governs development. Such values can be incorporated through markets, in which case, these values are reflected through voluntary exchanges. In contrast, sustainable development often seeks the "right prices" to internalize third-party effects through regulations.[12]

A more extreme version requires "maintenance of resources," meaning that no resource stocks should be diminished. This goal is impossible to achieve, however, if there is to be any present consumption of nonrenewable resources. With a copper mine, for example, none of the mineral could be taken from the earth and converted into valuable capital goods because the stock of copper available for future generations would be diminished. The maintenance of resources ignores the possibility that consumption of nonrenewable resources may reduce the consumption of other resources that are far more important to human life or the ecosystem. For example, by consuming petroleum, which is in finite but unknown supply, we can produce medical supplies that improve and extend human life. Does it make sense to save oil for future generations simply to guarantee them the same oil supply to which we have access? As economists Barnett and Morse concluded many years ago in their seminal study of resource scarcity, "By devoting itself to improving the lot of the living . . . each generation . . . transmits a more productive world to those who follow."[13] Strict adherence to the maintenance-of-resources tenet of sustainability precludes this investment.

Related to maintenance of resources is the precautionary principle, which, simply put, means "better safe than sorry." The precautionary principle is used to defend strict regulation of greenhouse gases on the grounds that we cannot take a chance that warming will occur even in the face of scientific uncertainty. As Scarlett and Shaw point out, this type of folk wisdom seems to make common sense, but it can be dangerous as a policy because being safe has its costs.[14] For example, eliminating chlorination from water reduces exposure to a possible carcinogen, but it can increase the risk from other diseases. Similarly, banning pesticides can eliminate potential carcinogens, but it also reduces agricultural productivity and raises the cost of food. We now know that automobile air bags save lives, but they also kill smaller people. And so on.

In short, the seemingly simple concept of sustainable development gets considerably more complex when we recognize opportunity costs and attempt to implement policy. If ecological principles and environmental ethics are to be factored into development policy, we still must ask who will do the factoring. Again, there is diversity of opinion among sustainable development advocates, but generally it is acknowledged that some "institutional modifications" will be necessary. A leading natural resource textbook summarizes these modifications.

1. An institution for stabilizing population
2. An institution for stabilizing the stock of physical wealth and throughput
3. An institution to ensure that the stocks and flows are allocated fairly among the population[15]

When these institutional modifications are dissected, the "beguiling simplicity and apparent self-evident meaning" of sustainable development are replaced with the reality of political controls to discipline the citizens.

> Fundamentally, "sustainable development" is a notion of . . . disciplining our current consumption. This sense of "intergenerational responsibility" is a new political principle, a virtue that must now guide economic growth. The industrial world has already used so much of the planet's ecological capital that the sustainability of the future is in doubt. That can't continue.[16]

The method of discipline is the primary distinguishing factor between sustainable development and free market environmentalism. Market prices discipline consumers to allocate their scarce budgets among competing demands, and they discipline producers to conserve scarcer, higher-priced resources by finding substitutes that are less scarce. This discipline works well as long as consumers and producers are faced with the costs of their actions. It breaks down, however, when property rights are ill-defined and ill-enforced. Sustainable development and free market environmentalism come together on the point that environmental problems arise when this discipline is lacking, but they diverge dramatically on what form this discipline should take. Advocates of sustainable development want political regulations to discipline markets, but free market environmentalists believe that market forces and property rights hold the key to managing the global commons.

GETTING THE SIGNALS RIGHT

If there are problems with discipline in the global commons, free market environmentalism teaches us that incentives matter and therefore that we should remove the perverse incentives that may make the problems worse or that induce people to ignore the consequences of their actions.

In the case of global warming causing a rise in ocean levels and changes in regional climates, it is important for individual decision makers to have good information and to be rewarded for acting appropriately. In other words, if global warming is occurring, individuals must not be sheltered from its consequences. For example, rising sea levels will inundate beaches, damaging people and property. Subsidies to beachfront development in the form of insurance and infrastructure reduce the costs of locating to where the effects of global warming will be greatest.[17] Removing these subsidies will not guarantee that beachfront development will not occur, but at least it will force individuals to accurately assess the risks. Advocates of regulation may argue that individuals will not have the knowledge to make the correct decisions, but with subsidies they have neither the knowledge nor the incentive to refrain from development.

Similarly, if global warming caused increased rainfall in some regions and decreased rainfall in others, then agricultural responses could range from abandoning farming in dry areas to developing drought-resistant crops. Unfortunately, current farm programs send the wrong signals. Subsidized irrigation and crop insurance encourages farmers to break prairie sod and plant crops in arid regions. Rather than choosing drought-resistant crops that might be more appropriate in an environment undergoing global warming, farmers intensify the use of pesticides and chemical fertilizers to increase their yields of price-supported crops grown on restricted acreages. Again, there is no guarantee that individual farmers will get it right, but current agricultural programs mask reality.

Another benefit of letting individuals respond to signals that reflect the consequences of global warming is that many people will experiment with how to respond to the problem. Rather than relying on a single building code or a specific drought-resistant crop, individual responses will create many potential solutions. Those that prove to be failures will be dropped in favor of those that are successful.

Political distortions have also contributed to the loss of tropical rain forests, which store enormous amounts of carbon dioxide and help regulate the buildup of carbon dioxide in the atmosphere. In Brazil, the government has expanded cattle ranching in the Amazon by offering subsidies and tax incentives; as a result, cattle ranching accounted for 72 percent of the 12,365,000 hectares altered by 1980.[18] Agricultural settlement, which has also been promoted by government investments in land settlement programs and by road building, has been the second most important cause of tropical rain forest destruction in Brazil. Other government investments have brought even more destruction. The Tucurui hydroelectric project on the Tocantis River, for example, was built at a cost of about $4 billion and has flooded 2,160 square kilometers of forestland.[19]

The failure of government to provide secure property rights and its encouragement of violation of property rights through land reform policies also discourage resource stewardship. Robert Deacon has examined many developing countries and found that government policies and governmental instability are the main contributors to deforestation rates.[20] Lee Alston, Gary Libecap, and Bernardo Mueller show how land reform in Brazil encourages conflict over property rights and discourages efficient land use.[21]

One of the more egregious examples of political distortions are subsidies to fishing fleets around the world. These subsidies can take a variety of forms, including subsidized loans for purchasing new vessels and income supports for fishers. They exacerbate the problem of the commons in fisheries in three ways. First, by artificially lowering costs and inflating profits, government subsidies attract more entrants into already financially strained fisheries. Second, by hiding the true cost of capital, government subsidies lead to overinvestments and higher fishing costs in the long run. Third, by propping up incomes, government subsidies encourage less efficient fishers to remain in fisheries that are being depleted from overfishing.

In a number of cases, fishing subsidies have paved the way for the collapse of a fishery. Recall the collapse of the cod fishery of Atlantic Canada discussed in chapter 5. In addition, the European Union (EU) financed a major fleet buildup and modernization for its member countries during the 1970s and 1980s. The dramatic fall of harvests in the 1980s is evidence that this subsidy stimulated fishing effort far in excess of what could be sustained. Current funding by the EU of member fishing fleets indicates its inability to learn from past mistakes. For the 1994–1999 period, the EU budgeted $747.7 million in "fleet renewal and modernization."[22]

Because many subsidy programs for fishing are either not budgeted (e.g., subsidized lending) or provided to support industries (e.g., harbors and shipyards), estimating the total amount of worldwide fishing subsidies is difficult. In 1992, the Food and Agriculture Organization of the United Nations estimated that most of the global fishing industry's annual deficit of $54 billion in the late 1980s was covered by government subsidies.[23] While the exact figure can be debated, this estimate gives some idea of the potential magnitude of fishing subsidies during the period. A more recent estimate from a 1998 World Bank study puts annual global fishing subsidies in the range of $15 billion to $21 billion, or about 20 to 25 percent of annual revenues from global ex-vessel fish sales.[24] Not included in these estimates are subsidies provided by subnational governments such as U.S. states and Chinese provinces.

During the Washington, DC, protests in April 2000 against international financing policies, the World Bank tried to portray a greener image. Unfortunately, its actions speak louder than its words. In Brazil and China, the bank continues to play the role of dam builder; in Indonesia, it continues to promote resettlement to pristine tropical rain forests; and in Thailand, its support for the state power system has wiped out river ecosystems. The World Bank has increased its funding of nongovernmental organizations with the noticeable result that these organizations are far less critical of the bank's actions.[25] The subsidized destruction of the environment caused by the World Bank and other international development institutions is perhaps unmatched in world history. The title of Bruce Rich's book says it all, *Mortgaging the Earth*.[26]

THE PROPERTY RIGHTS APPROACH

Given the intractability of national or international regulation, finding a solution to the global commons problem will require more innovative responses, including ways to define property rights and enforce them through the rule of law. Throughout this book, we have emphasized how property rights evolve and argued that government can encourage the establishment of property rights such as individual transferable quotas (ITQs) in fisheries. This property rights approach has the potential to yield truly innovative solutions to the tragedies of the global commons.

The first implication of the property rights approach is that countries with secure property rights and the rule of law enjoy greater economic growth than other countries. For example, using an index of property rights based on several variables, Gwartney, Lawson, and Block found that the growth rate in per capita gross domestic product (GDP) between 1980 and 1994 for countries with the highest security of property rights averaged 2.4 percent compared to –1.3 percent for countries with the lowest security index.[27] Similarly, Beach and Davis found that annual per capita GDP growth between 1980 and 1993 averaged 2.88 percent for countries with more secure property rights, compared with –1.44 percent for those with the least secure property rights.[28]

The relationship between these growth figures and environmental quality is important—higher incomes afford us more environmental quality in addition to material goods. It is no accident that less developed countries have more pollution, lower health standards, and more environmental hazards. The simple fact is that dynamic, growing economies, like dynamic ecosystems, are more resilient in coping with unanticipated environmental problems. Ecologist William Clark has pointed out that resilience is the essence of a healthy ecosystem: "The decreased frequency of variation in the system [is] accompanied by increased vulnerability to and cost of variation. . . ."[29] The late Aaron Wildavsky contrasted anticipation with resilience, stressing that we are better off avoiding the obvious, high-probability dangers and developing the resilience to deal with harms as they arise.[30]

Advocates of sustainable development argue that human betterment should be measured in terms of health, education, improved living standards for the most disadvantaged, and a cleaner environment. These conditions are precisely the results of economic growth. Countries with higher per capita incomes have better education, lower mortality rates, longer lives, and better living standards for the poor. If sustainable development is to mean no or even slower growth, third-world countries will not only be deprived of higher material living standards, they will also be deprived of the other measures of well-being, including health, safety, and education.

Seth Norton systematically measures the link between more secure property rights and measures of environmental quality.[31] Norton measures access to safe water and finds that 90 percent of the population in nations with strong property rights has access to safe water, while only 60 percent does in nations with weak property rights. As for sanitation, 86 percent of the population has access to sanitation in countries with strong property rights, compared with 53 percent in countries with weak property rights. Life expectancy is 15 to 20 years greater in countries with strong property rights. Countries with strong property rights saw forestlands increase slightly each year, while countries with weak property rights had an average annual reduction of forestlands of 1.52 percent. Norton concludes that "environmental quality and economic growth rates are greater in regimes where property rights are well defined than in regimes where property rights are poorly defined."[32]

The trick, of course, is to move from systems of weak property rights to those of strong ones, a problem with which economists have grappled since Adam Smith. In some cases, the best thing we can do is stay out of the way of property rights evolution. As seen in chapter 11, community norms can provide effective means for restricting access to the commons and encouraging good stewardship. Though economists have often equated common property or communal systems with open access, customs and culture can play a crucial role in limiting access to resources and encouraging good stewardship.[33] The evolution of property rights from the ground up often emanates from customary rules that are eventually codified and molded into property rights that can be exchanged in the marketplace. Too often, top-down regulations ignore these local institutions or make them illegal.

Private law can also play a role in enforcing property rights at the international level.[34] Indeed, without international common law enforcing property and contracts, debt-for-nature swaps and emissions trading programs would have little hope of succeeding. Perhaps one of the most famous cases of common law enforcing the right to be free from pollution is the *Trail Smelter* case. This dispute was between the Cominco Company in Canada, which allegedly spewed fumes that crossed the border, and the harmed cattle ranchers in the United States. The ranchers petitioned the U.S. government, and the case was taken to arbitration and settled. The arbitrators concluded that

> under the principles of international law, as well as the law of the United States, no State has the right to use or permit the use of its territory in such a manner as to cause injury by fumes in or to the territory of another or the properties or persons therein, when the case is of serious consequence and the injury is established by clear and convincing evidence.[35]

Hence, the ranchers were granted an injunction and awarded damages from Cominco. Certainly, common-law remedies will not work in all circumstances, but they should not be ignored as one element of establishing and enforcing property rights across borders.

And, of course, the evolution of property rights can be encouraged by governmental policies that establish property rights from the top down. As discussed in chapter 9, ITQs can help prevent overfishing even in open oceans, and establishing tradeable pollution permits such as transferable carbon emission permits can be a step in the direction of property rights and markets. The CAMPFIRE program in Zimbabwe is also an example of top-down property rights, as the central government gave control of wildlife to local councils and hence provided the incentive to manage wildlife on a sustainable basis. Unfortunately, recent events in Zimbabwe underscore the pitfalls of this approach. The same government that supported CAMPFIRE is now encouraging squatters to move onto private property, some of which are home to wildlife being managed on a sustainable basis. This may provide a test of what happens with insecure property rights.[36]

The evolution of property rights becomes more likely as new technologies make definition and enforcement feasible.[37] New electronic and satellite technologies make it feasible to track fishing boats so that trespassers can be detected and to track whales as they migrate. DNA testing makes it possible to identify specific animal populations. Satellites can monitor pollution on a real-time basis and can track the flows of pollution in the atmosphere as well as in water. Just as barbed wire revolutionized the ability of frontier settlers to define and defend the property rights to land, so can electronic, genetic, and satellite technologies move us toward global property rights solutions if the political institutions can be kept out of the way.

CONCLUSION

The suggestion that there may be a tragedy of the global commons is often accepted as a trump card supporting international regulation under the guise of kinder, gentler sustainable development. Often these regulations, however, call for limits on economic growth, leaving disproportionate burdens between developed and developing countries. In the absence of growth, those at the bottom of the economic ladder can only improve their lot by taking from those at the top, so population must be controlled, consumption must be curtailed, risks must be limited, new environmental ethics must be developed, and wealth must be redistributed.

Sustainable development implies that a regulatory disciplinary mechanism is required in addition to, if not in lieu of, markets. This usually means that there must be omniscient, benevolent experts who can model ecosystems and economies and determine the correct solutions. In order to attain the appropriate technology, the correct level of population growth, or the proper environmental ethic, political managers must have the necessary information, knowledge, and ethics to manage for sustainability. They must possess technical knowledge about quantities and qualities of resources, both human and physical, and they must have knowledge about what constitutes the material needs of both present and future generations. Furthermore, they must set aside any self-interest to manage for the benefit of present and future generations.

Because this form of scientific management is inimical to ecological and market processes, some advocates of sustainable development urge "social adaptation as the appropriate response to new ecological awareness, rather than more sophisticated expert-dominated management."[38] How this "societal adaptation" will come about is unclear, but the following Green Party manifesto emphasizes the role of government: "A Green Government would replace the false gods [of markets, greed, consumption, and growth] with cooperation, self-sufficiency, sharing and thrift."[39] This personification of a "Green Government" as the ecologically sensitive decision maker assumes that "the government," whether democratic or authoritarian, will be omniscient, benevolent, and ecologically sensitive. Accepting that the "Green Government" will take over the responsibility of guaranteeing ecologically sensitive development assumes that a political process is in place to channel the self-interest of voters and politicians toward that end. Certainly democratic processes can help channel governmental regulations in the direction of environmental quality, but we have seen ample evidence throughout this book that political action is not necessarily any more environmentally friendly than market processes.

In contrast, free market environmentalism is an approach to environmental problems that is consistent with the principles of ecology. Free market environmentalism accepts that individuals are unlikely to set aside self-interest and asks how institutions can harness this survival trait to solve problems. It recognizes that information about the environment is so diffuse that a group of experts cannot manage the planet as an ecosystem. Individuals must be relied upon to process time-and-place-specific information and to discover niches, just as other species do in the ecosystem.

Free market environmentalism emphasizes that economic growth and environmental quality are not incompatible. In fact, higher incomes afford us more environmental quality in addition to material goods. It is no accident that less developed countries have more pollution, lower health standards, and more environmental hazards. The simple fact is that dynamic, growing economies, like dynamic ecosystems, are more resilient in coping with unanticipated environmental problems.

Free market environmentalism emphasizes the importance of institutions that facilitate rather than discourage the evolution of property rights that are the basis of individual rights. Even if regulatory solutions can improve environmental quality, these benefits must be traded off against negative impacts on a dynamic economy and against the costs to individual freedom and liberty.

CHAPTER 13

PURITY VERSUS PRAGMATISM

Given government's less than stellar record at protecting the environment, it is hard to put faith in proposals that call for more governmental regulation to ensure environmental quality. Consider the environmental damage caused by governmental central planning in the socialist countries of Eastern Europe and in the Soviet Union prior to their collapse. Economist Mikhail Bernstam estimated that firms in these economies discharged more than twice as much air pollution as firms in Western market economies did.[1] Socialistic firms also used three times the energy to produce an equivalent amount of goods produced in Western market economies.[2]

Of course, Western democratic governments also have their failings when it comes to environmental protection, and the U.S. federal government is a prime example. The Environmental Protection Agency (EPA) recently named U.S. government facilities the nation's worst polluters of waterways, finding them out of compliance with Clean Water Act regulations more often than private facilities and local governments. Equally disturbing, government statistics "show a dramatic rise in federal government pollution of lakes and rivers since 1993."[3]

In addition, U.S. federal land, water, and wildlife managers have fared little better as resource stewards. The Bureau of Reclamation's and the Army Corps of Engineers' water projects led to the damming of nearly all major river systems in the western United States. A number of these projects continue to wreak havoc on legendary salmon and steelhead runs in the Pacific Northwest. The Forest Service loses hundreds of millions of dollars each year managing recreation, grazing, and timber on our national forests. At the same time, it allows forest health to deteriorate and fire risks to heighten by failing to use timber harvests and controlled burns to manage forest growth.[4] In the 1990s, the National Park Service added 3.4 million acres to its domain comprising 25 new parks of questionable value to its mission, yet it complains about the backlog of infrastructure projects such as roads, buildings, and sewage systems for which it needs funding.[5] These examples of "bureaucracy versus environment" illustrate that governmental command and control offers no guarantee of good natural resource stewardship or environmental quality.[6]

It is one thing to criticize governmental solutions to environmental problems, but it is quite another to go beyond the status quo to find pragmatic alternatives. Throughout this book, we have proposed property rights solutions that link self-interest with responsible use of environmental assets. In some cases, such as water marketing, strengthening the role of private ownership is a small step. In others, such as air pollution, however, the step toward property rights solutions is much greater. Establishing private property rights sometimes can be accomplished through customs and community norms, as shown in chapters 3 and 11, but more often there is a role for government as an institution that reduces the costs of definition and enforcement. Hence, one pragmatic approach toward free market environmentalism is to find ways that the government can facilitate the evolution of private property rights.

In cases where governmental ownership or regulation is firmly entrenched, however, privatization will be more difficult, if not impossible, thus calling for more incremental approaches. The tenets of free market environmentalism can provide a guide for pragmatic institutional changes that (1) get the incentives right for decision makers in the public and private sectors, (2) generate accurate information on environmental demands through market transactions, and (3) strengthen private property rights wherever possible. The efficacy of this strategy is evident in the strides forward that free market environmentalism made in the past decade.

THE POWER OF POSITIVE INCENTIVES

Though political solutions have created large federal bureaucracies, complex lawsuits, and large environmental organizations in Washington, DC, to lobby and litigate, they have not always delivered environmental progress at the grassroots level. Those impatient with the political approach, especially at the national level, increasingly are turning toward free market environmentalism to achieve their goals. For example, *USA Today* reports that "environmentalists across the country are offering financial incentives to farmers and ranchers to protect land and aid endangered animals."[7] As examples of this approach, *USA Today* points to the Nature Conservancy paying farmers in Indiana to help them purchase equipment needed for low-erosion tillage, and to the Environmental Defense Fund paying ranchers in Texas to trap and remove cowbirds that invade the nests of rare songbirds.

The growing list of entrepreneurial pragmatists in the environmental movement includes Hank Fischer, Northern Rockies director of Defenders of Wildlife. Having fought in the "wolf wars,"[8] Hank searched for a better alternative and came up with an insurance-like strategy to protect both wolves and ranchers.[9] Wolves are viewed by ranchers in the same way that environmentalists view pollution; wolves reintroduced into Yellowstone National Park spill out of its borders and cause harm to ranchers when they kill an occasional cow or sheep. To deal with this pollution problem, Defenders of Wildlife established its privately funded wolf compensation fund and offered to pay the ranchers for livestock losses due to wolf predation. Between its inception in 1987 and April 2000, the fund has paid compensation totaling $109,476.77 to 111 ranchers in the area surrounding Yellowstone National Park, central Idaho, and northwestern Montana.[10]

Clearly this insurance-type scheme is not perfect because it does not pay the rancher for the time spent proving the animal was killed by wolves or for the cost of actually purchasing a replacement animal. According to Margaret Soulen Hinson, a sheep rancher in Idaho who was paid from the fund for sheep lost to wolves, "We would rather have no losses than compensation."[11]

But given the highly emotional climate surrounding wolf reintroduction, imperfection is not surprising. And because ranchers' historic means of excluding wolves was to kill them, a system that compensates for the conflicting use of habitat—ranchers wanting land for grazing and environmentalists wanting it for wolves—moves wolf policy in the direction of negotiated settlements. Even Margaret Hinson recognizes that "it does make you more tolerant to participate in this program."

To further improve the program, Defenders of Wildlife established another fund to reward ranchers for allowing wolves to live on their private property. Hank Fischer recognized that the compensation program, at best, made ranchers neutral toward wolves; that is to say, with compensation, ranchers are not being asked to bear the full cost. By offering a reward to any rancher who has a wolf raise a litter of pups on private property, Defenders is trying to change the incentives. In the spring of 1994, a rancher near Augusta, Montana, collected a $5,000 reward from Defenders for having three wolf pups successfully raised on his property. The rancher told his cowboys to leave the wolves alone following advice from state and federal biologists about how to minimize human disturbance. By offering a reward to a rancher who allows the raising of wolves on his land, Defenders may be able to turn the liability of providing a public good into an asset.[12]

The compensation and reward programs are playing another important role by providing information about wolf predation and hence are laying the foundation for more private insurance schemes. For example, the programs provide an incentive for ranchers to report losses rather than shoot, shovel, and shut up, and it demonstrates how little damage there actually is. Moreover, because Defenders must pay compensation, it has an incentive to find ways to reduce the potential conflict over habitat. This program could eventually allow a private insurer to conduct actuarial studies to determine the best combination of premiums, risks, and payouts to satisfy both parties.

Another environmental organization using a pragmatic free market approach is the Delta Waterfowl Foundation, a private, nonprofit organization dedicated to helping the recovery of North American duck populations by providing breeding habitat in the United States and Canada. The foundation's flagship program is Adopt-a-Pothole, initiated in 1991. Funds are raised from private contributors, each of whom receives an aerial photograph of his or her adopted pothole, a quarterly report on its status, and an annual estimate of duck production. The farmer receives $7 an acre to maintain pothole habitat and $30 an acre to restore pothole habitat. Another innovation is the production contract. Delta Waterfowl offers farmers production contracts that pay them on the basis of how many ducks are produced on a given site. Production contracts give farmers an incentive to come up with innovative ways to increase the productivity of sites using their knowledge of the land and other local conditions.[13]

Results have been impressive. After nine years of operation, contributions to these efforts totaled over $3.6 million, and the list of adopted sites grew from 40 in 1991 to over 3,400 in 1999.[14] Nesting sites for mallards, canvasbacks, shovelers, blue-winged teal, green-winged teal, gadwalls, redheads, and pintails on farms in Manitoba, Minnesota, and North Dakota are fast becoming North America's duck factories. For instance, nest density is twice as great for adopted sites as compared with unadopted sites, and nesting success averages 51 percent for adopted sites, compared with 10 to 15 percent for unadopted sites. Moreover, the program has led to the development of a special nesting device that protects ducks from predators. Potholes using the device have nesting success rates of 90 percent.[15]

Elsewhere, a number of environmental groups are discovering the pragmatic approaches to public land management that build on the principles of free market environmentalism. "Gridlock" is a term that has aptly described traditional federal land policy, in which environmentalists are pitted against commodity users in a never-ending battle over how resources will be used.[16] As discussed in chapter 7, the gridlock is a manifestation of the political process, in which rights to use public lands are left up for grabs and players compete in a winner-take-all game that breeds conflict rather than cooperation.

If certain lands in the United States are going to be held in the public domain, then a pragmatic way of breaking the political gridlock is to rely more on market transactions to allocate land use rights. Such a strategy has recently caught the fancy of a number of environmental groups that have decided it is time to step up to the counter with cash in hand for the right to decide how assets on federal lands will be used. The first breakthrough came in the mid-1990s on a section of state school trust land in the state of Washington. Notably, school trust lands differ from federal lands in that they are required to generate income for public schools from timber, minerals, grazing, and recreation.[17]

Abandoning court battles and protests, a coalition of environmental groups bought the right not to cut timber on 25,000 acres of Loomis State Forest in north-central Washington. One thing that makes the Loomis deal significant is the size of the payment—$13.1 million—which was raised from over 5,000 private sources.[18] The deal marks the first time that private funds have been used to change how a government forest is managed.

The deal began to take shape when the Northwest Ecosystem Alliance was suing the Washington Department of Natural Resources over management of the Loomis Forest. A way out of gridlock was reached when the state and the alliance negotiated a settlement in which the alliance agreed to drop its suit and not challenge management of the Loomis if the department would accept the market value of the timber to be placed in a 25,000-acre Natural Resource

Conservation Area. With the payment made, the land will remain open to recreationists, and locals are guaranteed access for traditional uses, including off-road vehicles.

Mitch Friedman, director of the Northwest Ecosystem Alliance, one of five groups involved in the agreement, proclaims the purchase as "an exceptional opportunity for the people of Washington state."[19] The land they intend to prevent logging on is the last roadless area in the two-million-acre Loomis State Forest. The lodgepole pine forest has stunning mountains, clear streams, and abundant wildlife.

Repeatedly, the Northwest Ecosystem Alliance and other environmental groups have gone to court to block timber sales on state trust lands, but the new agreement ends the lawsuits, spares an area from logging, and provides income to Washington's school trusts. The coalition will pay the trust for the current generation of trees. In addition, it will pay for the purchase of an additional tract of land, elsewhere in the state, and give it to the state to be managed for timber production. Jennifer Belcher, Washington's commissioner of public lands, calls it a "real positive step" toward meeting both school funding and conservation goals.[20]

The power of the environmental purse is making itself felt on state and federal lands in other states as well. In 1996, the Forest Guardians of New Mexico, an environmental group known for its propensity to sue, outbid ranchers for a 644-acre state grazing lease in an area in which riparian habitat had been trampled down and the vegetation removed by cattle. Once they obtained the lease, the Forest Guardians removed the livestock and planted willows and other vegetation along the stream banks to provide wildlife habitat. While the total amount of money expended was small—only $770—the impact of that sale could affect the way millions of acres of public land are used in the future, simply by demonstrating that the market can allocate public lands.[21]

In 1999, Utah's Grand Canyon Trust announced it had organized a deal between ranching families and the Bureau of Land Management (BLM) to retire or relocate grazing allotments on about 120,000 acres inside the Grand Staircase-Escalante National Monument. Donations from individuals and foundations will pay the ranchers to retire their allotments. The price was not revealed to the public, but Bill Heddon of the trust said it was "well within fair market value."[22] A year earlier, the trust had made a similar arrangement to retire allotments in the Lost Spring Canyon addition to Arches National Park. Two years earlier, the Conservation Foundation also worked a deal to retire allotments in a section of Horseshoe Canyon west of Canyonlands National Park.

As long as the rancher is a willing seller, such buyouts provide a viable alternative to gridlock on public lands. Unfortunately, there may be a fly in the ointment that could stymie such an approach on federal lands. According to Jeff Burgess, a grazing-rights activist in Arizona, "Right now, if a third party came in and sued the BLM for the right to graze on these allotments in the monument, they would probably have to reissue the grazing permits."[23] Burgess, the Santa Fe Forest Guardians, and the Southwest Center for Biological Diversity of Tucson are part of a group that has asked the federal government for a guarantee that once an area has been retired from grazing (or another use) through voluntary exchange, federal land agencies cannot reissue the grazing permits to other grazers. Such clarification would go a long way toward encouraging similar brokerage deals on other federal lands. Unfortunately, on Washington's Okanogan National Forest, the Forest Service rejected a bid by the Northwest Ecosystem Alliance on a timber sale because the alliance wanted to carry out its option of letting the trees stand.[24]

Environmental groups willing to put "pragmatism over ideology"[25] are also making a difference in water allocation. Coho and chinook salmon are threatened by low water flows on the Little Applegate River southwest of Medford, Oregon, where more than 40 irrigators divert water from the river and sometimes dry it up. To make matters worse, a dam blocks fish from reaching the upper reaches of the river to spawn. The Oregon Water Trust and others recently helped organize a complex deal to get more water into the river for the fish. Through this deal, irrigators will dismantle the dam and give their water rights to the state. In return, they will get rights to water from another source and will get $1 million in private funding to transport the new water to their farms. The fish will get an additional 4,500 gallons per minute of stream flow.

CHANGING INSTITUTIONS

Sometimes implementing free market environmentalism requires changing the legal institutions that govern resource use. Though property rights may be well defined and enforced, transfers to willing buyers who place a higher value on the resource than the current owner may be precluded by legal constraints.

Water

An example of such a legal constraint is found with water rights that are well defined and enforced under the West's prior appropriation doctrine, but that are not easily transferred under legal and administrative rules. Hence, even though environmental groups may be willing to pay farmers more to leave the water instream than the water is worth if diverted, such willing buyer–willing seller transactions are not allowed. Typical political solutions, including reserving unclaimed water for fish or regulating use under the public trust doctrine, have promoted tremendous conflict without providing much water for fish.[26]

To allow markets to work to save streams, market-oriented groups first had to get the laws and administrative procedures changed. Such changes in Oregon, Washington, California, Idaho, and Montana have paved the way for impressive increases in water acquisitions for instream use, especially endangered species habitat.[27] Figure 13.1 charts the upward trend in water acquisitions for instream use through lease, purchase, and donation for 11 western states combined, from 1990 through 1997. Since 1990, an estimated $61 million has been spent on leases and purchases of water for instream use over this period. The market witnessed a significant jump in 1992. This increase reflected several state and federal acquisition programs—most notably, the San Joaquin Refuge water acquisition program and the New Interstate Stream Commission acquisition program prompted by the Pecos River Compact.

Figure 13.1
Annual Instream Flow Acquisitions, 1990–1997

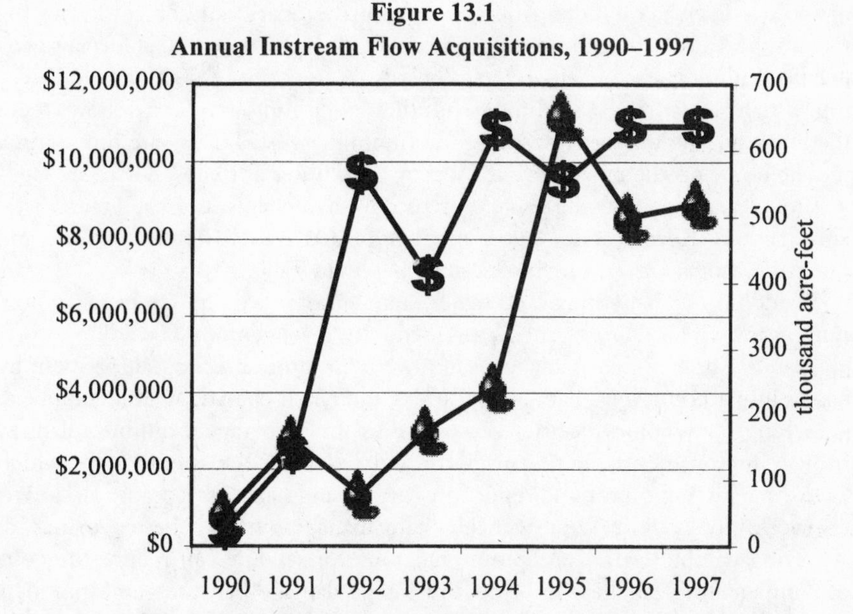

Source: Clay J. Landry, "Market Transfers of Water for Environmental Protection in the Western United States," working paper, WP98-3. Political Economy Research Center, Bozeman, MT, 1998.

Complementing the institutional changes at the state level are changes at the federal level that transfer ownership of federal water projects to local water users. Local water users assume responsibility for project operation in return for the right to trade surplus water to other users such as thirsty cities. As of April 1999, Congress had passed two pieces of legislation designating water transfers to several district water users in Texas and Idaho.[28]

Wildlife

Institutional change has also come to wildlife management in the western United States, where state wildlife agencies and landowners are forming partnerships to improve wildlife habitat. Under the ranching-for-wildlife approach, landowners typically promise to make their land more attractive to wildlife. They may develop new water sources, improve forage, enclose livestock, and plant trees along streams. In return for the improvements, landowners usually receive an extended hunting season and hunting permits that they can sell directly to hunters at whatever price the market will bear.

On the Prather Ranch, owner Steven Kerns manages 6,521 acres in Siskiyou County, California, for livestock and wildlife. It is high desert country surrounded by mountains. Under the state's ranching-for-wildlife program, the California Department of Fish and Game allows Kerns ten buck deer tags and one buck antelope tag to sell to hunters. Kerns sells the tags for $800 each, regardless of species. Each tag entitles a hunter to hunt the quarry of choice during the ranch's special two-month hunting season. This compares with California's general deer hunting season, which typically lasts about two weeks, and antelope season, which lasts a few days. Hunters run their own hunts, but Kerns shows them the ranch and points them toward good hunting areas. Hunters are housed in a mobile home on the ranch with full electric and water hookups. There are facilities for cleaning and storing game. In return for extended seasons and ranch-specific harvests, Kerns carries out a number of activities to enhance game and nongame habitat on the ranch, described below in Table 13.1.

In addition to benefiting landowners, ranching for wildlife opens up opportunities for state wildlife agencies to extend the management of wildlife to private lands. For example, managers can now attain more precise management by taking into account local habitat conditions and ranch-specific wildlife numbers. Ranching for wildlife specifies harvest limits for each ranch, outlines planned habitat improvements on the property, and describes the methods by which landowners will monitor and report on wildlife and harvests. Because landowners work closely with state agencies, state managers have a better chance of achieving wildlife goals. In addition, ranching for wildlife can reduce the costs of compensating ranchers for crop and forage damage because game numbers are better controlled. It can also take the place of expensive land purchases or cost-sharing programs to improve habitat.

Following successful experiments, California and Colorado legislatures made ranching for wildlife permanent in the 1980s. In the 1990s, ranching for wildlife has spread to other states based on growing satisfaction among landowners, hunters, and wildlife managers. As of 2000, eight states had ranching-for-wildlife programs. Four states—California, Colorado, Utah, and New Mexico—have comprehensive programs, while four others—Oklahoma, Washington, Nevada, and Oregon have fledgling programs. The remaining western states do not have

Table 13.1 Prather Ranch, Siskiyou County, California, Typical Habitat Improvements

Authorized Harvest:	10 buck deer, 1 buck antelope
•	Maintain a 5-acre cattle exclosure for fawning habitat
•	Maintain water levels in the two marsh areas to facilitate waterfowl brood survival
•	Plant 200 willow cuttings along Prather Creek
•	Make sure that no livestock grazing will occur on U.S. Forest Service allotment in 1998
•	Maintain eight permanent and four portable perch poles to improve Swainson's hawk habitat
•	Maintain 1,000 acres of irrigated alfalfa
•	Eliminate cattle grazing along Prather Creek
•	Maintain six burrowing-owl nest boxes

Source: Donald R. Leal and J. Bishop Grewell, *Hunting for Habitat: A Practical Guide to State-Landowner Partnerships* (Bozeman, MT: Political Economy Research Center, 1999), 61.

ranching for wildlife but do have landowner programs that could provide the basis for such partnerships in the future. Tables 13.2 and 13.3 below summarize the characteristics of the programs.

A fundamental goal of ranching for wildlife is to improve the environment for many kinds of wildlife, not just game animals. Environmentalists increasingly recognize that if landowners are to maintain open space and natural landscapes, they must have positive incentives. "As human populations continue to rise, so do market pressures to convert land into subdivisions and shopping malls," says Alan Christensen of the Rocky Mountain Elk Foundation. "If conservation groups do not step forward to work with private landowners, other interests will."[29]

Similarly, the Greater Yellowstone Coalition proposes ranching for wildlife as one way of protecting natural habitat around Yellowstone National Park. "Most open private lands that have not been developed belong to farmers and ranchers,"[30] says the coalition. Incentive-based programs like ranching for wildlife "increase the profitability of maintaining working ranches as open space and wildlife habitat."[31] Aldo Leopold, the grandfather of modern conservation thought, wrote that "conservation will ultimately boil down to rewarding the private landowner who conserves the public interest."[32]

Public Lands

Institutional changes for water and wildlife have come comparatively easy because they already involve some private property rights, but fostering similar institutional change for public lands is much tougher. Though full-fledged free market environmentalism would call for privatizing public lands to encourage better ecological and economic stewardship,[33] privatization is not politically palatable to either liberals or conservatives. This became obvious to the Reagan administration in 1982 when it recommended the sale of 35 million acres, or 5 percent, of the federal estate. This recommendation was the most visible effort at outright divestiture of government property since privatization under homesteading ended in the 1920s. But despite the modest amount of land to be sold,

Table 13.2 Ranches and Acres Enrolled

State	Number of Units	Number of Acres
California	55	685,378
Colorado	24	900,920
Nevada	—	—
New Mexico	1,442	~8,000,000
Oklahoma	75	250,000
Oregon	—	—
Utah	66	1,069,652
Washington	3	200,000

Source: Donald R. Leal and J. Bishop Grewell, *Hunting for Habitat: A Practical Guide to State-Landowner Partnerships* (Bozeman, MT: Political Economy Research Center, 1999), 17.

Table 13.3 Characteristics of State Programs

State Program	Date Established	Extended Seasons	Hunter Permits for Landowner	Minimum Acreage	Wildlife Management Plan Required	Habitat Improvement Required	Public Access Required
California	1984	yes	yes	none	yes	yes	no
Colorado	1989	yes	yes	12,000	yes	yes	yes
Nevada	1998	see note	yes	none	no	no	yes
New Mexico	see note	no	yes	none	no	no	yes
Oklahoma	1992	see note	yes	1,000	no	no	yes
Oregon	1995	no	yes	40	no	no	no
Utah	1994	yes	yes	10,000	yes	no	yes
Washington	1997	yes	yes	5,000	yes	yes	yes

Notes: The Nevada program allows some flexibility in when the hunter can go on the land. New Mexico's program evolved from state recognition of landowners who helped reintroduce elk into the state in the early 1900s. Oklahoma extends the season minimally to increase doe removal.

Source: Donald R. Leal and J. Bishop Grewell, *Hunting for Habitat: A Practical Guide to State-Landowner Partnerships* (Bozeman, MT: Political Economy Research Center, 1999), 18.

the disposal program was unceremoniously canceled only a year later. Those with hopes for the program failed to anticipate the opposition from not only environmentalists, but also from commodity interests.

Short of privatization, there have been two institutional experiments that have moved federal land management in the direction of free market environmentalism. One involves relying more on user-fee funding to pay for operation and maintenance of national and state lands. As noted in chapter 6, the 1996 Recreational Fee Demonstration Program is helping site managers improve customer service and pay for repairs and maintenance. But this program does not go nearly as far as states have in injecting market discipline into management of their state parks. Revenue raised from user fees in national parks still represents only a little over 10 percent of the National Park Service's $1.3 billion annual budget. The rest of the funding comes from taxpayers.

State park systems around the country are much more dependent on user fees for operating support. Since 1980, many state park systems have felt the pinch of fiscally tight legislatures. General tax support for state parks rose from $619 million in 1980 to $637 million in 1994. This is a small increase, given that consumer inflation rose by 74 percent over this period. As general support lagged, state park managers began to rely more on user fees. In 1980, fees collected at all state parks totaled $181.7 million, or about 17 percent of total park spending. Fourteen years later, in 1994, user fees totaled $637.9 million, or about 33 percent of total park spending. Notably, 16 park systems regularly obtain more than half their operating costs from user fees.[34]

New Hampshire has been a pioneer in operating self-sufficient parks. In April 1991, amidst a growing general-funds crisis, the state legislature passed an act requiring the park system to finance its operating budget through internally generated funds. The change in funding was cushioned by the fact that park income had actually exceeded operating expenditures for the three prior years. However, park receipts had been handed over to the treasury, breaking the direct link that provides a critical incentive for park managers to maintain parks in top condition to attract paying customers. The 1991 act restored that link by establishing a park fund to receive park earnings. The fund is dedicated to parks, and monies are carried over from year to year. This funding structure provides assurance to park personnel that the money is available to the parks and is also an incentive for them to maximize revenues. Innovation has been key to New Hampshire's success at earning a large portion of its operating budget. It was the first park system to implement differential pricing for campsites, taking into account the levels of amenities and of popularity of the sites. In addition, the park system was one of the first to institute per person entrance fees instead of charging by vehicle regardless of the number of passengers.[35]

The other institutional experiment involves applying a trust approach to certain federal lands. One variation of the trust approach to federal land management was recommended by Richard Stroup and John Baden in 1982.[36] They proposed establishing "wilderness endowment boards" that would be bounded by the common-law doctrine of trust to preserve wilderness areas. Each board would comprise representatives from environmental groups nominated by the

President and approved by Congress. Unlike the current approach to funding wilderness maintenance, these boards would cover the costs of wilderness maintenance out of revenues earned from wilderness assets such as oil and gas and other private sources.

Since that original trust proposal, the trust idea has been applied to a federal site. In the Omnibus Parks and Public Lands Management Act of 1996, Congress created a trust to manage the Presidio, a former military post that overlooks San Francisco's Golden Gate Bridge. The Presidio was the oldest continually operated military post in the nation. When it was decommissioned as an Army post, it was transferred to the National Park Service and became part of the Golden Gate National Recreation Area. An annual budget estimated to be as much as $38 million a year would have made the Presidio the costliest park in the park system. Several members of Congress complained that, were it turned into a typical park, the area would forever require subsidies to finance its upkeep. A trust approach instead was proposed whereby the Presidio would be funded from its endowment of man-made and natural resources, with the goal of becoming financially self-sufficient by 2013.[37]

The trust is responsible for managing the assets of the Presidio in a way that will minimize costs to the U.S. Treasury while preserving and enhancing amenities. Trust objectives include finding tenants and establishing programs to preserve the natural, historical, and cultural resources, while providing educational and recreational opportunities. The Presidio can be a community that promotes the ecological integrity of the site, socioeconomic diversity, and economic viability. The trust board of directors includes a designee of the secretary of the interior and six presidential appointees.

Unlike the managers of traditional parks, the Presidio board has a fiduciary responsibility to generate revenues by leasing its buildings and using its property in ways that will eventually cover all expenses. The board may use the revenues for administration, preservation, restoration, operation and maintenance, improvement, repair, and related expenses.

The Presidio, however, was not required to be financially self-sufficient immediately. It was given a budget of up to $25 million per year for up to 15 years. If the self-sufficiency goal cannot be attained after 15 years, all property under the Presidio's jurisdiction will be offered for sale to other federal agencies, public bodies, and private enterprises.

Of course, this is not a good way of getting the incentives right for managing the site, whether it is a trust or not. Allocating $25 million per year to the site's budget severely weakens the incentive to keep costs within the revenues generated for the 15 years in which the huge subsidy is provided. Finally, the legislation creating the trust left an important loophole. The trustees may transfer any portion of the property that they consider "surplus" to the secretary of the interior. Such an option allows the board to shift unprofitable properties to the National Park Service. This may serve to improve the financial condition of the trust, but it means more money-losing operations for the Park Service.

Nonetheless, at least with a goal of self-sufficiency, trustees have a clear measure by which they can assess performance. To meet this goal, they will have to consider ways to generate revenues and to use those revenues to enhance the Presidio's amenities. In addition, they will have to choose land and resource uses that will cover costs.[38]

Decentralized Pollution Control

As discussed in chapter 10, North Carolina's Tar-Pamlico Association became the first water-pollution-control trading community in North America in 1992.[39] The association, which is in essence a broker, emerged as a public-sector firm that facilitates the reduction of transaction costs among member dischargers.

Significantly it was the EPA that paved the way for this local solution. It relaxed its strict point-source regulations, established performance standards for the sound, and let the association decide how those standards would be met. This revision left the association free to devise a system of tradeable permits and effluent fees that encouraged high efficiencies in meeting the overall standard. The system the association developed facilitates trade among member point-source polluters and, most notably, nonmember, nonpoint dischargers, namely farmers. The latter group has proved to be especially troublesome under command-and-control approaches. The EPA's rule relaxation represented a significant shift in pollution policy at the national level.

Moreover, a rights-based river basin association such as Tar-Pamlico illustrates that a pragmatic approach to pollution regulation produces a far superior outcome to standard command-and-control regulations. Based on such success, the EPA has been pushing for more river basin associations modeled after Tar-Pamlico. In a 1996 report on watershed-based trading, the agency points to a new direction in pollution policy:

> Trading is an innovative way for water quality agencies and community stakeholders to develop common-sense, cost-effective solutions for water quality problems in their watersheds. . . . Trading can allow communities to grow and prosper while retaining their commitment to water quality.[40]

This change gives economist Bruce Yandle some cause for optimism when looking at the prospects of decentralization in air quality down the road. He writes,

> Some will continue the costly and less-effective command and control of the past. Others will go with property rights and market forces. Still others will move to performance standards and third-party environmental audits. All will rely on continuous monitoring of emissions and contract enforcement. Citizens who prefer more environmental quality will vote with their feet. Local and state politicians will respond by searching for lower-cost ways to provide more-effective environmental protection. Eventually, the country will have the equivalent of environmental enterprise zones, where centralized control gives way to market forces, property rights, and more-effective environmental management.[41]

CONCLUSION

During the 1990s, free market environmentalism caught the fancy of a growing number of environmentalists, who have turned to market solutions to solve environmental problems. Realizing that political environmentalism subjects resource management and environmental quality to the fickle whims of legislators and voters, some have turned to free market environmentalism because of its ability to provide a long-lasting solution. When property rights are well defined and enforced, environmentalists can be satisfied knowing that protection of a wetland or bird sanctuary is secured by common law precedent and by the Constitution of the United States. As the Sand County Foundation's president, Brent Haglund, put it when asked about the effectiveness of government programs in protecting wildlife habitat, "You know what I like? A deed in the courthouse."[42] Others have turned to market solutions simply because nothing else has worked. "It's very hard to make progress through lawsuits,"[43] says Bill Heddon of Grand Canyon Trust.

Yet others have turned to free market environmentalism because of its penchant for innovation. Aldo Leopold found property rights suitable for putting his nontraditional ideas to work on his Sand County Wisconsin farm, and that approach is still carried out today at the Leopold Memorial Reserve.[44] Greg Simonds and other wildlife managers use "holistic" range management to produce abundant wildlife on Utah's Deseret Ranch.[45] Similarly, Defenders of Wildlife and Delta Waterfowl are using creative solutions to overcome barriers to wolf reintroduction and waterfowl production on private lands.

On the policy front, free market environmentalism is making itself felt with institutional changes necessary to allow market transactions that can enhance environmental quality. Not just in the United States, but around the world, water marketing is improving water-use efficiency and enhancing environmental quality.[46] Changes in wildlife management on private lands in the United States pale in comparison to the role that markets play in preserving wildlife in southern Africa. Especially in South Africa, where game belongs to the landowner once property is fenced, marginal cattle ranching lands are being returned to wildlife habitat and private owners are becoming the salvation of endangered species such as the black rhinoceros.[47]

Public lands are experiencing change from business as usual. Subsidized recreation may be on the wane as user fees provide a direct link between consumers and land managers, and public land trusts give land managers more incentive to weigh tradeoffs between various land uses. Again, Africa and other developing countries are paving the way for self-sufficiency in park management simply because the governments of those countries do not have the capacity to adequately fund parks and protected areas.[48]

On the pollution front, free market environmentalism has made less progress mainly because of institutional barriers. Progress has been made in using pollution-permit trading schemes, but real property rights solutions have not been forthcoming. Some would say that this is because the costs of defining and enforcing property rights are simply too high to make this a feasible option.[49] But others contend that the problem is national regulations that stand in the way of common law solutions.[50] If there is reason for optimism in the pollution area, it is because the rising cost of achieving higher levels of air or water quality will require harnessing the entrepreneurial spirit of the marketplace, as has happened with Tar-Pamlico Sound.

We have come a long way down the free market environmental path, but there is still a long way to travel. The federal government controls one-third of the land in the United States—that amount has been growing by 800,000 acres per year since 1960 and is likely to grow even faster with congressional legislation that will commit billions of dollars to buying more land from private owners. The Endangered Species Act has been up for reauthorization for several years, with everyone recognizing that it has not been effective at saving species, but no one is willing to stand tall for real reform. The momentum of agencies such as the U.S. Fish and Wildlife Service, the EPA, and the Bureau of Reclamation is difficult to stop.

When the idea of free market environmentalism was launched in the 1980s, it was considered an oxymoron; today, however, there is hardly an environmentalist who would not concede that markets have a role to play in advancing environmental quality. The general tendency around the world toward freer markets means that populations of the developing world will share in the wealth that markets create and will have the wherewithal to join the ranks of environmentalists. Combining the spark of innovative ideas with the fuel of pragmatic entrepreneurship gives us hope that we can break the regulatory fist of command and control and replace it with a greener invisible hand.

NOTES

CHAPTER 1

1. These are quoted in Ronald Bailey, "Earth Day, Then and Now," *Reason On Line,* website: http://www.reson.com/0005/fe.rb.earth.html.
2. Donnella H. Meadows, Dennis L. Meadows, Jorgen Randers, William W. Behrens III, *The Limits to Growth: A Report for the Club of Rome's Project on the Predicament of Mankind* (New York: A Potomac Associates Book, New American Library, 1974), ix–x.
3. For a discussion of additional apocalyptic predictions, see Edith Efron, *The Apocalyptics* (New York: Simon and Schuster, 1984), chapter 1.
4. *Global 2000 Report to the President* (Washington, DC: U.S. Government Printing Office, 1980), 1. For a critique of the Global 2000 findings and for a date refuting the predictions, see Julian Simon and Herman Kahn, *The Resourceful Earth: A Response to Global 2000* (Oxford, England: Basil Blackwell, 1984).
5. Lester R. Brown, "The Future of Growth," in *State of the World 1998,* Lester R. Brown, project director (New York: W. W. Norton & Company, 1998), 16.
6. Ibid., 17.
7. For an excellent antidote to the Worldwatch predictions, see Ronald Bailey, ed., *Earth Report 2000: Revisiting the True State of the Planet* (New York: McGraw-Hill, 2000).
8. Reto Florin, chief of FAO's Water Resources, Development, and Management Service, at the Second World Water Forum, The Hague, March 17-22, 2000. See Food and Agriculture Organization of the United Nations, "Water to Feed the World: Perspectives for the Future," news and highlights, March 22, 2000, website: http://www.fao.org/news/2000/000306-e.htm.
9. Julian Simon, *The Ultimate Resource* (Princeton, NJ: Princeton University Press, 1981).
10. Paul Ehrlich, *The Population Bomb* (New York: Sierra Club-Ballantine Books, 1968).
11. Sandra Postel, "Facing Water Scarcity," in *State of the World 1993,* Lester R. Brown, project director (New York: W. W. Norton & Company, 1993), 40.
12. Mikhail S. Bernstam, "Comparative Trends in Resource Use and Pollution in Market and Socialist Economies," in *The State of Humanity,* ed. Julian Simon (Cambridge, MA: Blackwell Publishers, 1995), 503–22.
13. Randal O'Toole, "Learning the Lessons of the 1980s," *Forest Watch* 10 (January–February 1990): 6.

CHAPTER 2

1. John Maynard Keynes, *The General Theory of Employment, Interest, and Money* (New York: Harcourt, Brace & World, 1964), 383.
2. A. C. Pigou, *The Economics of Welfare* (London, England: Macmillan, 1920).
3. Ibid., 195.
4. Ibid.
5. Francis Bator, "The Simple Analytics of Welfare Maximization," *American Economic Review* (March 1957): 22–59.
6. Alston Chase, *Playing God in Yellowstone* (Boston: Atlantic Monthly Press, 1986).
7. John M. Hartwick and Nancy D. Olewiler, *The Economics of Natural Resource Use* (New York: Harper & Row, 1986), 18.
8. Daniel W. Bromley, *Property Rights and the Environment: Natural Resource Policy in Transition* (Cambridge, MA: Blackwell, 1991), 55.
9. Ronald Coase, "The Problem of Social Cost," *Journal of Law and Economics* 3 (October 1960): 1–44.
10. Garret Hardin, "The Tragedy of the Commons," *Science* 162 (December 1968).
11. Samuel P. Hays, *Conservation and the Gospel of Efficiency: The Progressive Conservation Movement, 1890–1920* (Cambridge, MA: Harvard University Press, 1959), 28.
12. Alan Randall, *Resource Economics: An Economic Approach to Natural Resource and Environmental Policy* (New York: Wiley, 1987), 36.
13. F. A. Hayek, "The Use of Knowledge in Society," *The American Economic Review* 35 (September 1945): 519–20.
14. Thomas Sowell, *A Conflict of Visions* (New York: William Morrow and Company, 1987), 46.
15. Hayek, "The Use of Knowledge," 520.
16. Ibid., 521–522.
17. Sowell, *A Conflict of Visions,* 48.
18. Hayek, "The Use of Knowledge," 521.
19. See James D. Gwartney and Richard L. Stroup, *Economics: Private and Public Choice,* 8th ed. (Orlando, FL: Harcourt Brace & Company, 1997), 785–809, for a discussion of the forces that lead to higher transaction costs in government.
20. Charles M. Tiebout, "A Pure Theory of Local Expenditures," *Journal of Political Economy* 64 (1956): 416–24.
21. See Donald R. Leal and Holly Lippke Fretwell, "Back to the Future to Save Our Parks," *PERC Policy Series* No. PS-10 (Bozeman, MT: Political Economy Research Center, June 1997), for a comparison of state and local parks.
22. Pamela Snyder and Jane S. Shaw, "PC Oil Drilling in a Wildlife Refuge," *Wall Street Journal,* September 7, 1995, A14.
23. Thomas Stratmann, "The Politics of Superfund," in *Political Environmentalism: Going Behind the Green Curtain,* ed. Terry L. Anderson (Stanford, CA: Hoover Institution Press, 2000).
24. See Terry L. Anderson and Peter J. Hill, *The Birth of a Transfer Society* (Stanford, CA: Hoover Institution Press, 1980), for a discussion of rent seeking.
25. Gwartney and Stroup, *Economics,* 785–809.
26. Anthony Fisher, *Resource and Environmental Economics* (New York: Cambridge University Press, 1981), 54.
27. Terry L. Anderson and Donald R. Leal, *Enviro-Capitalists: Doing Good While Doing Well* (Lanham, MD: Rowman and Littlefield Publishers, 1997).
28. For several articles critiquing free market environmentalism, see Mark Sagoff, "Free Market Versus Libertarian Environmentalism," *Critical Review* 6 (spring/summer 1992): 211–30.

29. Sagoff, "Free Market," 214.
30. The free market environmentalism argument is premised on the existence of property rights. It can always be argued that externalities exist and therefore that market exchanges won't work, but this is an efficiency argument, not a moral argument.
31. Sagoff, "Free Market," 218.
32. Coase, "The Problem of Social Cost."
33. Sagoff, "Free Market," 218.
34. Peter S. Menell, "Institutional Fantasylands: From Scientific Management to Free Market Environmentalism," *Harvard Journal of Law and Public Policy* 15 (1992): 489, 509.
35. Jane S. Shaw, "Environmental Regulation: How It Evolved and Where It is Headed," *Real Estate Issues* 1 (1996): 6.
36. Elinor Ostrom, *Governing the Commons: The Evolution of Institutions for Collective Action* (New York: Cambridge University Press, 1990).
37. Donald R. Leal, "Community-Run Fisheries: Avoiding the Tragedy of the Commons," *PERC Policy Series* No. PS-7 (Bozeman, MT: Political Economy Research Center, September 1996).
38. Terry L. Anderson, "Conservation—Native American Style," *PERC Policy Series* No. PS-6 (Bozeman, MT: Political Economy Research Center, July 1996).
39. Sagoff, "Free Market," 224.

CHAPTER 3

1. This chapter is adapted from Terry L. Anderson and Peter J. Hill, "From Free Grass to Fences: Transforming the Commons of the American West," in *Managing the Commons,* ed. Garrett Hardin and John Baden (San Francisco: W. H. Freeman, 1977), 200–16.
2. Eric Zuesse, "Love Canal: The Truth Seeps Out," *Reason* 12 (February 1981): 16–33.
3. Walter Prescott Webb, *The Great Plains* (New York: Grosset & Dunlap, 1931), 206.
4. Ibid., 17.
5. Ernest Staples Osgood, *The Day of the Cattleman* (Minneapolis: University of Minnesota Press, 1929), 182.
6. For a discussion of crowding on the open range, see ibid., 181–83.
7. Webb, *The Great Plains,* 229.
8. Quoted in Osgood, *Day of the Cattleman,* 183.
9. Ibid., 21, 201.
10. Maurice Frink, W. Turrentine Jackson, and Agnes Wright Spring, *When Grass Was King* (Boulder: University of Colorado Press, 1956), 98–99.
11. The stockgrowers' lobbying power declined dramatically because of the disastrous winter, and the 1889 territorial legislature repealed many stock laws. See W. Turrentine Jackson, "The Wyoming Stock Growers Association, Its Years of Temporary Decline, 1886–1890," *Agricultural History* 22 (October 1948): 265, 269.
12. Minutes of the Montana Stock Growers Association, 1885–1889, quoted by Ray H. Mattison, "The Hard Winter and the Range Cattle Business," *The Montana Magazine of History* 1 (October 1951): 18.
13. Alistair Cooke, *Alistair Cooke's America* (New York: Knopf, 1973), 237.
14. Jay Monaghan, ed., *The Book of the American West* (New York: Bonanza, 1963), 292.
15. Osgood, *Day of the Cattleman,* 193.
16. For a more complete description of the effort to claim public land, see Gary D. Libecap, *Locking Up the Range: Federal Land Control and Grazing* (San Francisco: Pacific Institute for Public Policy Research, 1981).
17. Osgood, *Day of the Cattleman,* 33, 114.

18. Frink et al., *When Grass Was King,* 12.
19. *Laws of the Montana Territory,* 1864–1865, sess. 1, 401; *Laws of Wyoming Territory,* 1869, sess. 1, chap. 62, 426–27.
20. Osgood, *Day of the Cattleman,* 124–26.
21. For a complete account of the use of barbed wire, see Webb, *The Great Plains,* 309.
22. U.S. Department of Agriculture, Bureau of Agricultural Economics, *Livestock on Farms, January 1, 1867–1935* (Washington, DC: U.S. Government Printing Office, 1938), 117.
23. For a more complete discussion of water rights, see Terry L. Anderson and Pamela S. Snyder, *Water Markets: Priming the Invisible Pump* (Washington, DC: Cato Institute, 1997).
24. Quoted in Webb, *The Great Plains,* 434.
25. Ibid., 433, 447.
26. Clesson S. Kinney, *Law of Irrigation and Water Rights and the Arid Region Doctrine of Appropriation of Waters* 1 (San Francisco: Bender-Moss, 1912), sec. 598.
27. Webb, *The Great Plains,* 444–48.
28. Wells A. Hutchins, *Water Rights Laws in the Nineteen Western States,* Miscellaneous Publication no. 1206, vol. 1 (Washington, DC: Natural Resources Economics Division, U.S. Department of Agriculture, 1971), 442–54.
29. Webb, *The Great Plains,* 446.
30. Gregory B. Christainsen and Brian C. Gothberg, "The Potential of High Technology for Establishing Tradeable Rights to Whales," paper presented at "The Technology of Property Rights," 1999 PERC Political Economy Forum, Bozeman, Montana, December 2–5, 1999.
31. Michael De Alessi, "Fishing for Solutions: The State of the World's Fisheries," in *Earth Report 2000: Revisiting the True State of the Planet,* ed. Ronald Bailey (New York: McGraw-Hill, 2000), and Christainsen and Gothberg, "The Potential of High Technology for Establishing Tradeable Rights to Whales."

CHAPTER 4

1. Douglass C. North, Terry L. Anderson, and Peter J. Hill, *Growth and Welfare in the American Past: A New Economic History* (Englewood Cliffs, NJ: Prentice-Hall, 1983), 111–21.
2. Theodore Roosevelt, in *Proceedings of the American Forest Congress* (Washington, DC: American Forestry Association, 1905), 9.
3. Gifford Pinchot, *The Fight for Conservation* (New York: Doubleday and Page, 1910), 123–24.
4. Robert F. Fries, *Empire in Pine: The Story of Lumber in Wisconsin* (Madison: State Historical Society of Wisconsin, 1951), 8–23, 250–51; Ronald N. Johnson and Gary D. Libecap, *Explorations in Economic History* 17 (1980): 376–77; Agnes M. Larson, *History of the White Pine Industry in Minnesota* (Minneapolis: University of Minnesota Press, 1949), 11, 29–28, 220–21, 404; Frederick Merk, *Economic History of Wisconsin During the Civil War Decade* (Madison: State Historical Society of Wisconsin, 1916), 60–73.
5. Sherry H. Olson, *The Depletion Myth: A History of Railroad Use of Timber* (Cambridge, MA: Harvard University Press, 1971).
6. Andrew D. Rodgers III, *Bernhard Edward Fernow: A Story of North American Forestry* (Princeton, NJ: Princeton University Press, 1951), 1.
7. Fries, *Empire in Pine,* 15.
8. Bernhard E. Fernow, *Economics of Forestry* (New York: Thomas Y. Crowell, 1902), 1.
9. Merk, *Economic History of Wisconsin,* 100, 105–8; Fries, *Empire in Pine,* 190, 245, 286–88. See also Paul W. Gates, *History of Public Land Law Development* (Washington,

DC: Public Land Law Review Commission, 1968), 534–55; Lucile Kane, "Federal Protection of Public Timber in the Upper Great Lakes States," *Agricultural History* 23 (1949): 135–39.

10. Russell McKee, "Tombstones of a Lost Forest," *Audubon* 90 (March 1988): 68.
11. In 1871, the average daily wage of a skilled laborer in the United States was $2.58, and good pine stands could be obtained for $4.00 an acre. See Bureau of the Census, *The Statistical History of the United States: From Colonial Times to the Present* (New York: Basic Books, 1976), 165; Paul W. Gates, *The Wisconsin Pine Lands of Cornell University: A Study in Land Policy and Absentee Ownership,* 2d ed. (Madison: State Historical Society of Wisconsin, 1965), 214. During the Civil War, "the wages of loggers in the northwestern pineries of Wisconsin ranged from $3 to $4 per day including board." See Merk, *Economic History of Wisconsin,* 109.
12. Merk, *Economic History of Wisconsin,* 108.
13. Ibid., 70.
14. See Oscar Burt and Ronald G. Cummings, "Production and Investment in Natural Resource Industries," *American Economic Review* 60 (1970): 576–90; Howard Hotelling, "The Economics of Exhaustible Resources," *Journal of Political Economy* 39 (1931): 137–75; Robert M. Solow, "The Economics of Resources or the Resources of Economics," *American Economic Review* 64 (May 1974): 1–14.
15. This assumes that costs are not rising or falling. If they are, the price could rise or fall faster than the interest rate. Technically, it is the rental value of the resource that follows the interest rate.
16. This same tragedy of the commons evident on the American frontier explains harvest practices in many developing regions such as Amazonia. See Lee J. Alston, Gary D. Libecap, and Bernardo Mueller, *Titles, Conflicts, and Land Use: The Development of Property Rights and Land Reform on the Brazilian Amazon Frontier* (Ann Arbor: University of Michigan Press, 1999).
17. Gates wrote that in 1852 "[t]he Territorial Legislature of Minnesota stated that encouragement had been given to the establishment of sawmills in the territory but not an acre of pine land had been offered at public sale and none was open to preemption." The legislature further stated that the industry "would be willing and anxious to pay the government for the land. . . ." See Gates, *Wisconsin Pine Lands,* 538.
18. This was not accidental. The government understood the importance of private ownership in the protection of the timber resource. See Fries, *Empire in Pine,* 192.
19. Johnson and Libecap, *Explorations in Economic History,* 379; Milton Friedman and Anna J. Schwartz, *A Monetary History of the United States* (Princeton, NJ: Princeton University Press, 1963), 69.
20. George F. Warren and Frank A. Pearson, *Prices* (New York: John Wiley and Sons, 1933), 36.
21. By the turn of the century, timber from the South and the Far West was beginning to dominate the market. See Johnson and Libecap, *Explorations in Economic History,* 376–77; Larson, *History of the White Pine Industry,* 221, 396–98.
22. For a detailed description, see Olson, *The Depletion Myth,* 42–69.
23. For a detailed discussion of how speculation can promote the optimal use of resources, see Terry L. Anderson and P. J. Hill, "The Race for Property Rights," *Journal of Law and Economics* 33 (April 1990): 177–97.
24. Richard N. Current, *Pine Logs and Politics: A Life of Philetus Sawyer, 1816–1900* (Madison: State Historical Society of Wisconsin, 1950), 22–25; Merk, *Economic History of Wisconsin,* 73; Johnson and Libecap, *Explorations in Economic History,* 375.
25. Current, *Pine Logs and Politics,* 23.
26. Gates, *Wisconsin Pine Lands,* 106, 237–39, 242, 243.

27. Information on the Kingston Plains is from McKee, "Tombstones of a Lost Forest," and from conversations with people in the Great Lakes area.
28. The $20 figure is based on Gates, *Wisconsin Pine Lands,* 238. Interest rate sources are Friedman and Schwartz, *A Monetary History,* 640, and *Economic Report of the President—February 1988* (Washington, DC: U.S. Government Printing Office, 1988), 330.
29. Warren Scoville, "Did Colonial Farmers 'Waste' Our Lands?" *Southern Economic Journal* 20 (1953): 178–81.
30. A. G. Ellis, "Upper Wisconsin Country," in *Collections of the State Historical Society of Wisconsin,* ed. Lyman C. Draper, vol. 3 (Madison: State Historical Society of Wisconsin, 1857), 445.
31. Larson, *History of the White Pine Industry,* 405.
32. Merk, *Economic History of Wisconsin,* 99.
33. Malcolm Rosholt, *The Wisconsin Logging Book* (Rosholt, WI: Rosholt House, 1980), 282.
34. The Porcupine Mountains Wilderness State Park in Minnesota is particularly interesting because it was nearly logged due to Department of Defense contracts during World War II. It should be noted that it was government demand rather than private market forces that finally made the timber valuable enough to consider for logging.
35. See Steven Karpiak, "The Establishment of Porcupine Mountains State Park," *Michigan Academician* 2 (1978): 135–39. For information on other tracts, see Fred Rydholm, "Upper Crust Camps," in *A Most Superior Land: Life in the Upper Peninsula of Michigan* (Lansing: Michigan Natural Resources Magazine, 1983). Additional information concerning the holdings of the Huron Mountain Club can be obtained from the Huron Mountain Wildlife Foundation in White Pigeon, Michigan.
36. Bill Ogden, telephone conversation, April 21, 2000. Family members on his wife's side have been members of the club for years.
37. Aldo Leopold, *Report on Huron Mountain Club* (Huron, MI: Huron Mountain Club, 1938), 40.
38. For a more complete discussion of the Huron Mountain Club, see Terry L. Anderson and Donald R. Leal, *Enviro-Capitalists: Doing Good While Doing Well* (Lanham, MD: Rowman and Littlefield Publishers, 1997), 30–33.
39. Quoted in Alfred Runte, *Trains of Discovery* (Niwot, CO: Roberts Rinehart, 1990), 23.
40. See Carlos Schwantes, *Railroad Signatures across the Pacific Northwest* (Seattle: University of Washington Press, 1993).

CHAPTER 5

1. Terry L. Anderson and Jane Shaw, "Grass Isn't Always Greener in a Public Park," *Wall Street Journal,* May 28, 1985, 30.
2. Ibid.
3. All dollar figures are in 1996 dollars.
4. Holly Lippke Fretwell, *Public Lands: The Price We Pay,* Public Lands Report No. 1 (Bozeman, MT: Political Economy Research Center, August 1998), 1, 12.
5. Donald R. Leal, "Turning A Profit On Public Forests," *PERC Policy Series* No. PS-4, (Bozeman, MT: Political Economy Research Center, September 1995), 4–5.
6. Fretwell, *Public Lands,* 6.
7. Ibid., 8–9.
8. Ibid., 16–17.
9. Robert F. Smith, chairman, *Statement,* Hearing to Review the Forest Service Timber Sale Program. Committee on Agriculture, U.S. House of Representatives. Photocopy, June 11, 1998.

10. Bill Schultz, *Forestry Best Management Practices Implementation Monitoring* (Missoula, MT: Montana Department of State Lands, 1992).
11. U.S. General Accounting Office (GAO), *Forest Service Decision-Making: A Framework for Improving Performance,* GAO/RECD-97-71 (Washington, DC, 1997), 32.
12. Steve Arno, "The Concept: Restoring Ecological Structure and Process in Ponderosa Pine Forests," in *The Use of Fire in Forest Restoration,* USFS Intermountain Research Station INT-6TR-341, June 1996, 37.
13. Michael Dombeck, Forest Service chief, congressional testimony, March 18, 1997.
14. National Park Service Organic Act, 16 U.S.C.1.
15. Michael Milstein, "Park Water Risky," *Billings* [Montana] *Gazette,* July 5, 1998.
16. "Yellowstone Closes Part of the Grand Loop Road," *Island Park News,* August 14, 1998.
17. James M. Ridenour, *The National Parks Compromised: Pork Barrel Politics and America's Treasures* (Merrillville, IN: ICS Books, 1994), 108.
18. John G. Mitchell, "Our National Parks: Legacy at Risk," *National Geographic* 186 (October 1994): 54.
19. "U.S. Park Service Spends $333,000 on Outhouse," *ENN Daily News,* October 8, 1997.
20. Edward T. Pound, "Costly Outhouses Monuments to Red Tape," *USA Today,* December 15, 1997.
21. Frank Greve, "Senior Legislators Claim Funds For Pet Projects," *Washington Post,* December 1, 1997.
22. Ibid.
23. Richard W. Wahl, *Markets for Federal Water: Subsidies, Property Rights, and the Bureau of Reclamation* (Washington, DC: Resources for the Future, 1989), 197–219; Kathleen Rude, "Ponded Poisons," *Ducks Unlimited* 54 (January–February 1990): 14–18.
24. Northwest Power Planning Council (NPPC), *Columbia River Basin Fish and Wildlife Program* (typescript copy), Portland, Oregon (February 1994): 2–13.
25. Idaho Department of Fish and Game (IDFG), "Can We Bring Back Salmon Fishing?" in *Saving Idaho's Salmon.* Pamphlet issued by IDFG, Boise, Idaho.
26. Andrew Herr, "Saving the Snake River Salmon: Are We Solving the Right Problem?" *PERC Working Paper* No. 94–15. Bozeman, MT: Political Economy Research Center, August 26, 1994.
27. Richard W. Wahl, "Cleaning Up Kesterson," *Resources* 83 (spring 1986): 12.
28. Elizabeth Brubaker, "Property Rights: Creating Incentives and Tools for Sustainable Fisheries Management," in *Fraser Forum* (BC, Canada: The Fraser Institute, April 1998): 10.
29. Ibid.
30. R. Quentin Grafton, "Performance of and Prospects for Rights-Based Fisheries Management in Atlantic Canada," in *Taking Ownership: Property Rights and Fishery Management on the Atlantic Coast,* ed. Brian Lee Crowley (Halifax, Nova Scotia, Canada: Atlantic Institute for Market Studies, 1996), 145.
31. Task Force on Atlantic Fisheries [Canada], *Navigating Troubled Waters for the Atlantic Fisheries,* report (Ottawa, Canada: Ministry of Supply and Services, 1982), 60.
32. C. A. Bishop, E. F. Murphy, M. B. Davis, J. W. Baird, and G. A. Rose, *An Assessment of the Cod Stock in NAFO Divisions 2j+3k,* NAFO Scientific Council Research Document 93/86, serial number N2771, 1993.
33. Price Waterhouse, *Human Resources Development Canada: Operational Review of the Atlantic Groundfish Strategy* (Ottawa, Canada: Price Waterhouse, 1995).
34. Brubaker, "Property Rights," 10.
35. Ibid.

36. Bruce Rich, *Mortgaging the Earth: The World Bank, Environmental Impoverishment, and the Crisis of Development* (Boston: Beacon Press, 1994), 7–8.
37. "Once It's Here . . . ," *The Economist*, March 1, 1997, 19–20.
38. Rich, *Mortgaging the Earth,* 77.
39. World Bank, Portfolio Management Task Force, *Effective Implementation: Key to Development Impact,* report (Washington, DC: World Bank, October 2, 1992), 4.
40. Ibid., iii, 4, 12.
41. Ibid.
42. Rich, *Mortgaging the Earth,* 27.
43. Ibid., 28.
44. Ibid.
45. Ibid.
46. Ibid., 28–29.
47. Ibid., 36.
48. Ibid.
49. Ibid.
50. Ibid., 36–37.

CHAPTER 6

1. Figures on recreational visits to the park provided by Don Striker, Comptroller of Yellowstone National Park, March 12, 1996.
2. Figures for 1960 are from U.S. Department of Commerce, Bureau of the Census, *Statistical Abstract of the United States, 1987* (Washington, DC, 1987), Table 380, 219; figures for 1996 are from U.S. Department of the Interior, Fish and Wildlife Service, and U.S. Department of Commerce, Bureau of the Census, *1996 National Survey of Fishing, Hunting, and Wildlife-Associated Recreation* (Washington, DC, November 1997), Table 12, 69, and Table 17, 74. Increases are adjusted for CPI increase from 1960 through 1996.
3. Kenneth E. Solomon, "South Dakota Fee Hunting: More Headaches or More Wildlife Problems on Agricultural Lands," ed. D. L. Hallett, W. R. Edwards, and G. V. Burger (Bloomington, IN: North Central Section of the Wildlife Society, 1988), 229–38; Jim Robbins, "Ranchers Finding Profit in Wildlife," *New York Times,* December 13, 1987; James P. Sterba, "Plight of the Pheasant Frames the Debate Over Hunting's Future," *Wall Street Journal,* February 1, 1999.
4. *1996 National Survey of Fishing, Hunting, and Wildlife-Associated Recreation,* 5.
5. Terry L. Anderson, "To Fee or Not to Fee: The Economics of Below-Cost Recreation," in *Multiple Conflicts Over Multiple Uses,* ed. Terry L. Anderson (Bozeman, MT: Political Economy Research Center, 1994), 4.
6. The exception is in the East and in the mid-South region, where federal land is less pervasive, and in the provision of facilities such as ski runs or campgrounds, where additional capital investment is necessary and the private sector has responded.
7. Reflecting this trend is *U.S. Hunting Report,* ed. Don Causey (Miami, FL: Oxpecker Enterprises). Since 1996, this monthly newsletter has reported on the growing number of ranches and farms that provide fee hunting opportunities for big game in the West.
8. These figures were obtained from National Wilderness Preservation System's home page, website: http://www.wilderness.net/nwps/; and the Interagency Wild and Scenic Rivers Coordinating Council and the National Trails System Map and Guide-Text, at: http://www.nps.gov/htdocs1/pub_aff/naltrail.htm, accessed February 11, 1998.
9. President's Commission on American Outdoors. American Outdoors: *The Legacy, The Challenge* (Washington, DC: Island Press, 1987), xi.

10. Terry L. Anderson, "Camped Out in Another Era," *Wall Street Journal,* January 14, 1987.
11. Lonnie L. Williamson, "Wildlife Superbill Reintroduced," *Outdoor News Bulletin,* January 29, 1999.
12. Traci Watson, "Clinton Pushes Preservation Plan," *USA Today,* January 12, 1999; Haya El Nasser, "Gore Veers From Past Growth Policy," *USA Today,* January 12, 1999.
13. Randall G. Holcombe, professor of economics at Florida State University, takes exception to this argument. He notes that "developed areas in the United States, excluding Alaska, are only 6.2 percent of the nation's total land area," far less pervasive than the federal government, which is the nation's largest landowner. See "Urban Sprawl: Pro and Con," *PERC Reports* 17(1) (Bozeman, MT: Political Economy Research Center, February 1999): 3–5.
14. Task Force on Recreation on Private Lands, *Recreation on Private Lands: Issues and Opportunities* (proceedings from a workshop sponsored by the President's Commission on Americans Outdoors, Washington, DC, March 10, 1986), 1.
15. Scott McMillion, "Gallatin Forest Is A Gift To Community That Costs Taxpayers," *Bozeman Daily Chronicle,* August 31, 1997.
16. In many situations, liability protection is considered a normal cost of the activity. Businesses elect to buy liability insurance because the expected costs of a lawsuit exceed the insurance costs. These insurance premiums become a cost of doing business and are reflected in prices. If consumers are not willing to pay the price, including insurance costs, then the product will not be supplied. If ranchers are precluded from charging for fishing on their property, then they are not adequately compensated for the increased liability and, therefore, they supply less recreational opportunities.
17. A new U.S. Geological Survey study entitled *Arctic National Wildlife Refuge, 1002 Area, Petroleum Assessment, 1998* puts the mean estimate for oil under the coastal plain at 20.7 billion barrels, or 50 percent higher than the previous estimate done in 1987. Website: http://energy.usgs.gov/factsheets/ANWR/ANWR.html.
18. Donald Woutat, "Stakes Are High in the Battle Over Exploration in Alaska National Wildlife Refuge," *Bozeman Daily Chronicle,* November 5, 1987.
19. Victor H. Ashe, "Needs and Opportunities for Outdoor Recreation," in *Transactions of the Fifty-first North American Wildlife and Natural Resources Conference,* ed. Richard E. McCabe (Washington, DC: Wildlife Management Institute, 1986), 14.
20. Harold Demsetz, "Toward a Theory of Property Rights," *American Economic Review* 57 (May 1967): 348.
21. For a more complete discussion of wildlife contracting problems, see Dean Lueck, "The Economic Organization of Wildlife Institutions," in *Wildlife in the Marketplace,* ed. Terry L. Anderson and Peter J. Hill (Lanham, MD: Rowman and Littlefield Publishers, 1995).
22. In the natural resource arena, the traditional arguments against the market provision of natural amenities have focused on public goods and common property. Public goods occur where existing property rights do not allow exclusion to capture the real demand for the good. Common property occurs where rights are held in common by a group of individuals, none of whom has transferable ownership interest. Access to the resource may be unrestricted, as in the case of air, or controlled politically, as most wildlife is controlled by state governments in the United States. These arguments do not go far enough in asking what obstacles stand in the way of establishing private property rights.
23. In October 1989, the Convention on International Trade in Endangered Species of Wild Fauna and Flora (CITES), meeting in Lausanne, Switzerland, voted to classify

the African elephant as an endangered species and to abolish the legal ivory market. The blanket ban on ivory trading prevailed until June 1997, when it was partially lifted, at the request of southern African countries, to allow limited sales of existing ivory stockpiles.

24. Urs P. Kreuter and Randy T. Simmons, "Economics, Politics and Controversy Over African Elephant Conservation," in *Elephants and Whales: Resources for Whom?* ed. Milton M. R. Freeman and Urs P. Kreuter (Switzerland: Gordon and Breach Science Publishers, 1994), 39–57.
25. Ibid., 49.
26. Ibid., 48–49.
27. "As the Cattle Business Weakens, Ranchers Turn Their Land Over to Recreational Use," *Wall Street Journal,* August 27, 1985, 33.
28. Tom Blood and John Baden, "Wildlife Habitat and Economic Institutions: Feast or Famine for Hunters and Game," *Western Wildlands* 10 (spring 1984): 13.
29. At the beginning of fiscal year 1997, 40 national forests were selected to participate in the Recreational Fee Demonstration Program. On these forests recreationists are usually charged a daily access fee and may be charged an activity fee.
30. Holly Lippke Fretwell, *Public Lands: The Price We Pay,* Public Lands Report No. 1 (Bozeman, MT: Political Economy Research Center, 1998), 6, 9, 12, and 17.
31. President's Council on Environmental Quality, *15th Annual Report of the Council on Environmental Quality* (Washington, DC: U.S. Government Printing Office, 1984), 426.
32. Letter from Richard A. Boitnott, Manager–Wildlife Ecology, International Paper Company, mid-South Region, Shreveport, Louisiana, March 8, 1994.
33. Data obtained from a 1998 brochure entitled *North Maine Woods.* Write: P.O. Box 421, Ashland, Maine 04732.
34. Ibid.
35. John S. Baen, "The Growing Importance and Value Implications of Recreational Hunting Leases to Agricultural Land Investors in America," presented at the American Real Estate Society meeting in Sarasota, Florida, April 16–19, 1997. A copy of the study is available from John S. Baen, Associate Professor of Real Estate, College of Business Administration, Finance, Insurance, Real Estate and Law, University of North Texas, Denton, Texas.
36. Don Causey, ed., "Found! A Season-Long Deer Hunt For Only $700!" *U.S. Hunting Report* (June 1997): 1.
37. Brochure on Fay Fly Fishing Properties, Inc., Bozeman, Montana.
38. Ibid.
39. Aldo Leopold, "Game and Wild Life Conservation [1932]," in *The River of the Mother of God and Other Essays by Aldo Leopold,* ed. Susan L. Flader and J. Baird Callicott (Madison: University of Wisconsin Press, 1991), 166.
40. "Private Clubs Provide Choice Shooting," *Fishing and Hunting News* 12 (April 1982): 2.
41. "Land Trusts Are Booming," *Bozeman Daily Chronicle,* December 4, 1998.
42. Gordon Abbott, Jr., "Long-Term Management: Problems and Opportunities," in *Private Options: Tools and Concepts for Land Conservation,* ed. Barbara Rusmore, Alexandra Swaney, and Allan D. Spader (Covello, CA.: Island Press, 1982), 207.
43. E-mail response from Daniel Yu, Member Service Center, The Nature Conservancy, March 9, 2000.
44. Sue E. Dodge, ed., *The Nature Conservancy Magazine* 40 (March–April 1990): 3, 33.
45. For details on this program, see Holly Lippke Fretwell, "Paying To Play: The Fee Demonstration Program," *PERC Policy Series* No. PS-17 (Bozeman, MT: Political Economy Research Center, December 1999).

46. James B. Coffin, "Clinton Administration Again Recommends Permanent Fee Demo," *Federal Parks & Recreation,* February 25, 2000, 8.
47. Michael Milstein, "GAO Report Praises Park Fee System," *Billings* [Montana] *Gazette,* February 16, 1999.
48. Don Causey, "This State's Landowner Program A Real Winner," *U.S. Hunting Report* (June 1998): 8–10.
49. For data on the impact of farm programs on duck habitat, see Daniel K. Benjamin, Kurtis J. Swope, and Terry L. Anderson, "Bucks for Ducks or Money for Nothin'? The Political Economy of the Federal Duck Stamp Program," in *Political Environmentalism: Going Behind the Green Curtain,* ed. Terry L. Anderson (Stanford, CA: Hoover Institution Press, 2000).
50. See the "angling statute,"§87-2–305, MCA, which recognizes a public right to fishing access up to the high watermark.
51. Helena AP, "FWP Commission rejects Ruby River closure request," *Bozeman Daily Chronicle,* February 5, 1999.
52. See, for example, Montana Wildlife Federation, "Public Ownership of Wildlife & the Threat of Privatization," Helena, Montana, February 22, 1999.
53. Donald R. Leal and J. Bishop Grewell, *Hunting for Habitat: A Practical Guide to State-Landowner Partnerships* (Bozeman, MT: Political Economy Research Center, 1999).
54. *Lake Shore Duck Club v. Lake View Duck Club,* 50 Utah 76, 309 (1917).
55. Eric Wiltse, "Irrigation Spells Death for Hundreds of Ruby River Trout," *Bozeman Daily Chronicle,* May 12, 1987.
56. Terry L. Anderson and Donald R. Leal, "A Private Fix for Leaky Trout Streams," *Fly Fisherman* 19 (June 1988): 28–31.
57. Dayton O. Hyde, "Recreation and Wildlife on Private Lands," in *Recreation on Private Lands,* published by the Task Force on Recreation on Private Lands. Proceedings of a workshop held in Washington, DC, March 10, 1986, 25.
58. Ibid., 26.
59. 16 U.S.C. § 1538(a)(1)(B) and 16 U.S.C. § 1532 (19).
60. Richard L. Stroup, "The Economics of Compensating Property Owners," *Contemporary Economic Policy* 15 (1997): 55–65.
61. Dean Lueck and Jeffrey Michael, "Preemptive Habitat Destruction Under the Endangered Species Act," unpublished manuscript, Department of Agricultural Economics and Economics, Montana State University, Bozeman, MT. Also see Dean Lueck, "The Law and Politics of Federal Wildlife Preservation," in *Political Environmentalism: Going Behind the Green Curtain,* ed. Terry L. Anderson (Stanford, CA: Hoover Institution Press, 2000).
62. Lueck and Michael, "Preemptive Habitat Destruction," 24.
63. Terry L. Anderson, Vernon L. Smith, and Emily Simmons, "How and Why to Privatize Federal Lands," *Cato Policy Analysis* 363, December 9, 1999.

CHAPTER 7

1. U.S. Geological Survey (USGS), *Arctic National Wildlife Refuge, 1002 Area, Petroleum Assessment, 1998,* assessment results, website: http://energy.usgs.gov/factsheets/ANWR/results.html.
2. Terry L. Anderson and Peter J. Hill, *The Birth of a Transfer Society* (Stanford, CA: Hoover Institution Press, 1980); Gordon Tullock, "The Welfare Costs of Tariffs, Monopolies, and Theft," *Western Economic Journal* 5 (June 1967): 224–32.

3. As of March 9, 2000, the amount of designated wilderness totaled 104,571,344 acres. See the National Wilderness Preservation System website: http://www.wilderness.net/nwps/.
4. U.S. General Accounting Office (GAO), *Federal Lands: Information on Land Owned and on Acreage with Conservation Restrictions,* GAO/RCED-95–73FS, Washington, DC, January 1995, 35.
5. American Petroleum Institute, "Access to Government Lands For Increased U.S. Oil Development," July 14, 1998, website: http://www.api.org/news/backup/496access.htm.
6. CNN, "Clinton Extends Moratorium On Offshore Oil Drilling" (June 1998), website: http://www-cgi.cnn.com/TECH/science/9806/12/offshore.drilling.pm.
7. U.S. Department of the Interior, Fish and Wildlife Service, *Draft Arctic National Wildlife Refuge, Alaska Coastal Plain Resource Assessment: Report and Recommendation to the Congress of the United States and Legislative Environmental Impact Statement,* November 1986.
8. The model used by the Wilderness Society used cost data based on a study published by the National Petroleum Council, *U.S. Arctic Oil and Gas* (Washington, DC: National Petroleum Council, 1981). A real oil-price growth rate of one percent from the 1987 base price of $18 per barrel is employed in half of the trails, and growth rates of zero and 2 percent each are used in 25 percent of the trails for the base-price scenarios.
9. W. Thomas Goerold, "Environmental and Petroleum Resource Conflicts: A Simulation Model to Determine the Benefits of Petroleum Production in the Arctic National Wildlife Refuge, Alaska," *Materials And Society* 11 (1987): 279–307.
10. Anderson and Hill, *Birth of a Transfer Society;* Tullock, "The Welfare Costs of Tariffs, Monopolies, and Theft."
11. For example, oil and gas leases on Montana's Gallatin and Flathead national forests were set aside in 1985 because a district court judge ruled that the environmental assessment carried out by the Forest Service was inadequate. As of March 1999, the status of these leases has not changed. According to Leslie Viculik, oil and gas specialist with the Forest Service, the lease acres amount to approximately 1.3 million acres between the two forests (e-mail from Earl Sutton, U.S. Forest Service, Northern Region, March 15, 1999).
12. 42 U.S.C. Sec. 4321.
13. Ed Porter and Lee Huskey, "The Regional Economic Effect of Federal OCS Leasing: The Case of Alaska," *Land Economics* 57 (November 1981): 594.
14. Richard Martin, "Resisting an Oil Rig Invasion," *Insight,* March 14, 1988, 17–18; Ken Wells, "U.S. Oil Leasing Plan Is Challenged by Eskimos Trying to Protect Their Culture at World's Edge," *Wall Street Journal,* March 12, 1986.
15. U.S. Department of the Interior, Minerals Management Service, *Mineral Revenues 1997: Report on Receipts from Federal And Indian Leases* (Denver: Minerals Management Services, 1997), Table 43, 122.
16. American Petroleum Institute, *Should Federal Onshore Oil and Gas Be Put Off Limits?* (Washington, DC: American Petroleum Institute, June 1984), 87.
17. Coastal Zone Management Act (as amended), section 307, 43 U.S.C. 1456.
18. 52 U.S.L.W. 4063, *Secretary of the Interior et al. v. California et al.* (U.S. January 11, 1984), 683 F. 2d 1253 reversed (9th Cir. 1982).
19. Porter and Huskey, "Regional Economic Effect," 594.
20. For example, section 311 of the Clean Air Act limits the liability of responsible parties to $50 million per incident of spillage unless willful negligence or willful misconduct on the part of the operator can be proven in a court of law.
21. James Everett Knight, Jr., "Effect of Hydrocarbon Development on Elk Movements and Distribution in Northern Michigan" (Ph.D. diss., University of Michigan, Ann Arbor, 1980).

22. American Petroleum Institute, *Compatibility of Oil and Gas Operations on Federal Onshore Lands with Environmental and Rural Community Values* (Washington, DC: American Petroleum Institute, 1984), 57; V. Van Ballenberghe, "Final Report on the Effects of the Trans-Alaska Pipeline on Moose Movements," Special Report 1 (Anchorage: Joint State/Federal Fish and Wildlife Advisory Team, 1976); Stering Eide and Miller Sterling, "Effects of the Trans-Alaska Pipeline on Moose Movements" (Juneau: Alaska Department of Fish and Game, June 1979).
23. The National Institute for Urban Wildlife, *Environmental Conservation and the Petroleum Industry* (Washington, DC: American Petroleum Institute, n.d.), 5.
24. Ken Baskin, "The Tug of War for the Wilderness," *Sun* (autumn 1985): 7.
25. H. Smets, "Compensation for Exceptional Environmental Damage Caused by Industrial Activities," in *Insuring and Managing Hazardous Risks: From Seveso to Bhopal and Beyond,* ed. Paul R. Kleindorfer and Howard C. Kunreuther (Berlin, Germany: Springer-Verlag, 1987), 80; Bill Richards, "Amoco Ordered to Pay Award of $85.2 Million," *Wall Street Journal,* January 12, 1988.
26. Harry Hurt III, Lynda Wright, and Pamela Abramson, "Alaska After," *Newsweek,* September 18, 1989, 50–62.
27. Exxon Valdez Oil Spill Trustee Council, "The Settlement," *Legacy of an Oil Spill: Ten Years After Exxon Valdez,* March 27, 2000, website: http://www.oilspill.state.ak.us/settlement/setlment.htm.
28. Two other platform accidents occurred in the Gulf of Mexico during 1970. One occurred in U.S. waters and the other occurred in Mexican waters. Neither resulted in significant amounts of oil reaching shore. See American Petroleum Institute, *Should Offshore Oil Be Put Off Limits?,* 119 and 131.
29. Walter J. Mead and Philip Sorenson, "The Economic Cost of Santa Barbara Oil Spill," in *Santa Barbara Oil Spill: An Environmental Inquiry* (Santa Barbara: California Marine Science Institute, University of California at Santa Barbara, 1972).
30. Los Angeles (CNN), "California oil spill endangers wildlife: headway made in cleanup," September 30, 1997, on-line version, http://www.cnn.com/US/9709/30/calif.oil.spill/; The Associated Press, "Workers combat 25 mile oil slick in Gulf of Mexico," October 3, 1998, on-line version, http://www.willjohnston.com/articles_98/october98/10_3_98wctosigom.html.
31. American Petroleum Institute, *Should Offshore Oil Be Put Off Limits?* 124–25, 134–35.
32. The National Institute for Urban Wildlife, *Environmental Conservation and the Petroleum Industry,* 13–15.
33. A study by the National Academy of Sciences estimated that world offshore oil and gas operations were responsible for 5 percent of the oil that gets into the world's oceans. The study noted that rivers were the principal source of oil pollution in the seas, accounting for 41 percent of the total. Tankers and other transportation forms account for 20 percent, natural oil seeps account for 15 percent, municipal and industrial effluent accounts for 11 percent, atmospheric sources (e.g., rain) account for 4 percent, urban runoff accounts for 3 percent, coastal refineries 1 percent, and U.S. offshore production just 0.05 percent. See U.S. Department of the Interior, Minerals Management Service, "Offshore Oil Production Accounts for Little of World's Ocean Pollution" (news release, Washington, DC, July 26, 1983).
34. For a complete discussion of the history of this production, see John Baden and Richard Stroup, "Saving the Wilderness: A Radical Proposal," *Reason* 13 (July 1981): 28–36. Also see Pamela Snyder and Jane S. Shaw, "PC Oil Drilling in a Wildlife Refuge," *Wall Street Journal,* September 7, 1995, A14.
35. John G. Mitchell, "The Oil Below," *Audubon* 83 (May 1981): 16–17.
36. Ibid.

37. See Marion Clawson, *The Federal Lands Revisited* (Washington, DC: Resources for the Future, 1983).
38. For one of the earliest proposals of this approach, see Richard L. Stroup and John A. Baden, "Endowment Areas: A Clearing in the Policy Wilderness," *Cato Journal* 2 (winter 1982): 691–708.
39. For details of this proposal, see Terry L. Anderson and Holly Lippke Fretwell, "A Trust for Grand Staircase-Escalante," *PERC Policy Series* No. PS-16 (Bozeman, MT: Political Economy Research Center, September 1999).

CHAPTER 8

1. Richard D. Lamm, "Forward," in *Western Water: Tuning the System,* by Bruce Driver (Denver: Western Governors' Association, 1986), ii.
2. Thomas J. Graff, "Future Water Plans Need a Trickle-Up Economizing," *Los Angeles Times,* June 14, 1982, V-2.
3. Rodney T. Smith, *Trading Water: The Legal and Economic Framework for Water Marketing* (Claremont, CA: Claremont McKenna College, Center for Study of Law Structures, 1986), 26.
4. Terry L. Anderson, "Institutional Underpinnings of the Water Crisis," *Cato Journal* 2 (winter 1983): 759–92.
5. Richard W. Wahl, *Markets for Federal Water: Subsidies, Property Rights, and the Bureau of Reclamation* (Washington, DC: Resources for the Future, 1989), 33.
6. Ibid., 27 and 46.
7. Delworth B. Gardner, *Plowing Ground in Washington: The Political Economy of U.S. Agriculture* (San Francisco: Pacific Research Institute for Public Policy, 1995), 298.
8. Wahl, *Markets for Federal Water,* 198–205.
9. Terry L. Anderson, "Water Options for the Blue Planet," in *The True State of the Planet,* ed. Ronald Bailey (New York: Free Press, 1995), 269.
10. Terry L. Anderson and Pamela S. Snyder, *Water Markets: Priming the Invisible Pump* (Washington, DC: Cato Institute, 1997), 104–5.
11. Ibid.
12. Terry L. Anderson and Pamela S. Snyder, "Priming the Invisible Pump," *PERC Policy Series* No. PS-9 (Bozeman, MT: Political Economy Research Center, February 1997): 15–16.
13. Testimony before the Subcommittee on Water and Power, House of Representatives, by the Honorable Eluid Martinez, Commissioner, Bureau of Reclamation, February 2, 1999.
14. Constitucion Politica de la Republica de Chile, chapter 3, article 24, final paragraph: "Los derechos de los particulares sobre las aguas, reconocidos o constituidos en conformidad a la ley, otorgaran a sus titulares la propiedad sobre ellos."
15. Renato Gazmuri Schleyer, "Chile's Market-Oriented Water Policy: Institutional Aspects and Achievements," in *Water Policy and Water Markets,* ed. Guy Le Moigne et al., World Bank Technical Paper 249 (Washington, DC: World Bank, 1994), 76.
16. Anderson and Snyder, *Water Markets,* 195.
17. The focus is on the Colorado River, where California and Nevada face shortages and high costs for alternative supplies and where Arizona is awash in subsidized water from the Central Arizona Project. See Anderson and Snyder, *Water Markets,* 197–98.
18. Quoted in *Water Intelligence Monthly,* "Babbitt Endorses Water Markets and Places High Premium on Political Consensus," January 1996, 2–3.
19. For a more detailed discussion of the "use it or lose it" principle, see Anderson and Snyder, *Water Markets,* 59, 87–88, and 115.
20. *Lake Shore Duck Club v. Lake View Duck Club,* 50 Utah 76, 166, 309 (1917).

21. 158 Colo. 331, 406 P. 2d 798 (1965).
22. *Fullerton v. California State Water Resources Control Board,* 90 Cal. App. 3d 590, 153 Cal. Rptr. 518 (1979); *California Trout, Inc., v. State Water Resources Control Board,* 90 Cal. App. Ed 816, 153 Cal. Rptr. 672 (1979).
23. R. W. Johnson, "Public Trust Protection for Stream Flows and Lake Levels," *University of California at Davis Law Review* 14 (1980): 256–57.
24. Driver, *Western Water,* 33–34.
25. Terry L. Anderson and Ronald N. Johnson, "The Problem of Instream Flows," *Economic Inquiry* 24 (October 1986): 535–54.
26. James Huffman, "Instream Uses: Public and Private Alternatives," in *Water Rights: Scarce Resource Allocation, Bureaucracy, and the Environment,* ed. Terry L. Anderson (San Francisco: Pacific Institute for Public Policy Research, 1983), 275.
27. *Dallas Times Herald,* July 16, 1984, B3.
28. Terry L. Anderson and Donald R. Leal, *Enviro-Capitalists: Doing Good While Doing Well* (Lanham, MD: Rowman and Littlefield Publishers, 1997), 98–99.
29. Douglas Southerland, *The Landowner* (London, England: Anthony Bond, 1968), 110.
30. Brian Clarke, "The Nymph in Still Water," in *The Masters of the Nymph,* ed. J. M. Migel and L. M. Wright (New York: Nick Lyons, 1979), 219.
31. Ed Zern, "By Yon Bonny Banks," *Field and Stream* 86 (September 1981): 120.
32. Clay J. Landry, *Saving Our Streams Through Water Markets: A Practical Guide* (Bozeman, MT: Political Economy Research Center, 1998), 7–8.
33. Ibid., 4.
34. Ibid., 30–31.
35. Kyra Epstein, "Water Trusts Provide Market-Driven Solution For Conservation, Water Quality," *U.S. Water News* 16 (May 1999): 21.
36. Richard W. Guldin, *An Analysis of the Water Situation in the United States: 1989–2040,* USDA Forest Service General Technical Report RM-177 (Washington, DC: U.S. Government Printing Office, 1989), 13.
37. Jean Margat, "A Hidden Asset," *UNESCO Courier* 15 (1993): 15.
38. Gregory S. Weber, "Twenty Years of Local Groundwater Export Legislation in California: Lessons from a Patchwork Quilt," *Natural Resources Journal* 34 (summer 1994): 660.
39. Terry L. Anderson and Pamela S. Snyder, "A Free Market Solution to Groundwater Allocation in Georgia," *Issue Analysis* (Atlanta: Georgia Public Policy Foundation, 1996); David Todd, "Common Resources, Private Rights and Liabilities: A Case Study on Texas Groundwater Law," *Natural Resources Journal* 32 (summer 1992): 233–63; and Guldin, *An Analysis,* 15.
40. Terry L. Anderson, Oscar Burt, and David Fractor, "Privatizing Groundwater Basins: A Model and Its Applications," in Huffman, *Water Rights.*
41. Frank J. Trelease, "Developments on Groundwater Law," in *Advances in Groundwater "Mining" in the Southwestern States,* ed. Z. A. Saleem (Minneapolis: American Water Resources Association, 1976), 272.
42. Terry L. Anderson and Peter J. Hill, "Privatizing the Commons: An Improvement?" *Southern Economics Journal* 50 (October 1983): 438–50.
43. A more elaborate modification has been proposed. See Jacque L. Emel, "Groundwater Rights: Definition and Transfer," *Natural Resources Journal* 27 (summer 1987): 653–73.
44. For a discussion in the context of oil, see Stephen N. Wiggins and Gary D. Libecap, "Oil Field Unitization: Contractual Failure in the Presence of Imperfect Information," *American Economic Review* 75 (June 1985): 370.
45. Adjudication of the Mojave Basin, located 100 miles east of Tehachapi Basin, provides a more recent example of improvements. See Anderson and Snyder, *Water Markets,* 182–186.
46. Anderson and Snyder, *Water Markets,* 178–182.

CHAPTER 9

1. For an example of pollution at sea, see Ronald Mitchell, "Intentional Oil Pollution of the Oceans," in *Institutions for the Earth: Sources of Effective International Protection,* ed. Peter M. Haas et al. (Cambridge, MA: MIT Press, 1993), 183–247. For two examples of stock depletion close to home, see Michael De Alessi, "Fishing for Solutions: The State of the World's Fisheries," in *Earth Report 2000: Revisiting the True State of the Planet,* ed. Ronald Bailey (New York: McGraw-Hill, 2000), 87–89.
2. Food and Agriculture Organization of the United Nations (FAO), *The State of World Fisheries and Aquaculture 1998,* on-line version, website: http://www.fao.org/docrep/w9900e/w9900e00.htm.
3. U.S. Department of the Interior, Fish and Wildlife Service, and U.S. Department of Commerce, Bureau of the Census, *1996 National Survey of Fishing, Hunting, and Wildlife-Associated Recreation* (Washington, DC: U.S. Department of Interior, November 1997); Nelson Bryant, "Fishing Licenses Are At Issue," *New York Times,* February 5, 1989; Gina Maranto, "Caught in Conflict," *Sea Frontiers* 35 (May–June 1988): 144–51; National Marine Fisheries Service (NMFS), *Our Living Oceans: Report on the Status of U.S. Living Marine Resources, 1999,* U.S. Department of Commerce, NOAA Technical Memorandum NMFS-F/SPO-41, on-line version, website: http://spo.nwr.noaa.gov/fa3.pdf.
4. Ross Eckert, *The Enclosure of Ocean Resources* (Stanford, CA: Hoover Institution Press, 1979), 4.
5. Ibid., 16.
6. National Marine Fisheries Service (NMFS), *Report to Congress: Status of Fisheries of the United States,* September 1998, on-line version, website: http://www.nmfs.gov/sfa/98stat.pdf.
7. National Marine Fisheries Service, *Our Living Oceans,* website: http://spo.nwr.noaa.gov/national.pdf; National Marine Fisheries Service (NMFS), *The Economic Status of U.S. Fisheries, 1996,* U.S. Department of Commerce, National Oceanic and Atmospheric Administration, National Marine Fisheries Service, December 1996, on-line version, website: http://www.st.nmfs.gov/econ/oleo/oleo.html.
8. The term "tragedy of the commons" was taken from Garrett Hardin, "The Tragedy of the Commons," *Science* 162 (December 1968): 1243–48.
9. For a classic article on the commons problem, see H. Scott Gordon, "The Economic Theory of a Common Property Resource: The Fishery," *Journal of Political Economy* 62 (April 1954): 124–42. See also Colin W. Clark, "Profit Maximization and the Extinction of Animal Species," *Journal of Political Economy* 81 (August 1981): 950–60.
10. Frederick W. Bell, "Technological Externalities and Common-Property Resources: An Empirical Study of the U.S. Northern Lobster Fishery," *Journal of Political Economy* 80 (January–February 1972): 156.
11. See Tom Tietenberg, *Environmental and Natural Resource Economics,* 2d ed. (Glenview, IL: Scott, Foresman and Company, 1988), 258–64.
12. J. A. Crutchfield and G. Pontecorvo, *The Pacific Salmon Fisheries: A Study of Irrational Conservation* (Baltimore: Johns Hopkins University Press, for Resources for the Future, 1969); "The Flaw in the Fisheries Bill," *Washington Post,* April 13, 1976.
13. Francis T. Christy, Jr., and Anthony Scott, *The Common Wealth in Ocean Fisheries* (Baltimore: Johns Hopkins University Press, for Resources for the Future, 1965), 15–16; Crutchfield and Pontecorvo, *The Pacific Salmon Fisheries,* 46.
14. James A. Crutchfield, "Resources from the Sea," in *Ocean Resources and Public Policy,* ed. T. S. English (Seattle: University of Washington Press, 1973), 115.

15. Robert Higgs, "Legally Induced Technical Regress in the Washington Salmon Fishery," *Research in Economic History* 7 (1982): 82.
16. Clarence G. Pautzke and Chris W. Oliver, *Development of Individual Fishing Quota Program for Sablefish and Halibut Longline Fisheries off Alaska* (Anchorage: North Pacific Fishery Management Council, 1977), 2.
17. Pub. L. No. 94–265, sec. 303(b)(6), 94th Congress, H.R. 200, April 13, 1976.
18. Anthony Scott, "Market Solutions to Open-Access, Commercial Fisheries Problems," paper presented at Association for Public Policy Analysis and Management 10th Annual Research Conference, Seattle, October 27–29, 1988, 7–8.
19. Ibid.
20. Maranto, "Caught in Conflict," 145.
21. William J. Chandler, ed., *Audubon Wildlife Report, 1988/1989* (San Diego: Academic Press, 1988), 48.
22. Based on an exchange rate of US$0.69 for every NZ$1.00. See Tom McClurg, "Bureaucratic Management versus Private Property: ITQs in New Zealand after Ten Years," in *Fish Or Cut Bait,* ed. Laura Jones and Michael Walker (Vancouver, BC: The Fraser Institute, 1997), 103.
23. Rodney P. Hide and Peter Ackroyd, "Depoliticising Fisheries Management: Chatham Islands' Paua (Abalone) as a Case Study," unpublished report for R. D. Beattie Ltd. Centre for Resource Management, Lincoln University, Christchurch, New Zealand, March 1990, 42, 44.
24. Michael De Alessi, *Fishing for Solutions* (London, England: The Institute of Economic Affairs, 1998), 43; Peter Hartley, *Conservation Strategies for New Zealand* (Wellington, New Zealand: New Zealand Business Roundtable, 1997), 97.
25. Ragnar Arnason, "Property Rights as an Organizational Framework in Fisheries: The Cases of Six Fishing Nations," in *Taking Ownership: Property Rights and Fishery Management on the Atlantic Coast,* ed. Brian Lee Crowley (Halifax, Nova Scotia: Atlantic Institute for Market Studies, 1996), 120–21.
26. As the phrase implies, individual vessel quotas allocate percentage shares of the total allowable catch to individual vessels. Like ITQs, these shares are transferable.
27. Christopher Sporer, "An Intelligent Tale of Fish Management," *Fraser Forum,* December 1998, 12–13.
28. William L. Robinson, "Individual Transferable Quotas in the Australian Southern Bluefin Tuna Fishery," in *Fishery Access Control Programs Worldwide: Proceedings of the Workshop on Management Options for the North Pacific Longline Fishers,* Alaska Sea Grant Report no. 86-4 (Orca Island, WA: University of Alaska, 1986), 189–205.
29. Arnason, "Property Rights as an Organizational Framework in Fisheries," 113.
30. ITQs are innovative yet controversial for the feared effects on communities and special interests that have invested in the status quo of strictly regulated fisheries. As of this writing, a moratorium on adopting ITQs in other federal fisheries has been imposed by Congress since 1996.
31. National Research Council, *Sharing the Fish: Toward a National Policy on Individual Fishing Quotas* (Washington, DC: National Academy Press, 1999), 293; National Marine Fisheries Service (NMFS), *Economic Status of U.S. Fisheries, 1996.*
32. National Research Council, *Sharing the Fish,* 312.
33. Parzival Copes, "A Critical Review of the Individual Quota as a Device in Fisheries Management," *Land Economics* 62 (August 1986): 278–91.
34. National Research Council, *Sharing the Fish,* 108.
35. Ibid., 108–110.
36. This occurred in the initial stages of Iceland's ITQ fisheries, but was later rectified. See National Research Council, *Sharing the Fish,* 329.

37. For complete discussions, see Terry L. Anderson and P. J. Hill, "Privatizing the Commons: An Improvement?" *Southern Economic Journal* 50 (1983): 438–50, and Terry L. Anderson and P. J. Hill, "The Race for Property Rights," *Journal of Law and Economics* 33 (April 1990): 177–97.
38. Hide and Ackroyd describe this problem in the context of New Zealand's efforts to establish ITQs. See Hide and Ackroyd, "Depoliticising Fisheries Management."
39. Higgs, "Legally Induced Technical Regress," 59.
40. Ibid.
41. Scott, "Market Solutions," 19.
42. Francis T. Christy, "Paradigm Lost: The Death Rattle of Open Access and the Advent of Property Rights to Regimes in Fisheries," paper prepared for the 8th Biennial Conference of the Institute of Fisheries Economics and Trade, Marrakesh, Morocco, July 1–4 ,1996, 14.
43. Richard J. Agnello and Lawrence P. Donnelley, "Prices and Property Rights in the Fisheries," *Southern Economic Journal* 42 (October 1979): 253–62.
44. James J. Acheson, "Capturing the Commons: Legal and Illegal Strategies," in *The Political Economy of Customs and Culture: Informal Solutions to the Commons Problem,* ed. Terry L. Anderson and Randy T. Simmons (Lanham, MD: Rowman and Littlefield Publishers, 1993), 69–83.
45. Donald R. Leal, "Community-Run Fisheries: Avoiding the Tragedy of the Commons," *PERC Policy Series* No. PS-7 (Bozeman, MT: Political Economy Research Center, September 1996).
46. Ibid., 13.
47. Ibid., 6–7.
48. Ronald N. Johnson and Gary D. Libecap, "Contracting Problems and Regulation: The Case of the Fishery," *American Economic Review* 12 (December 1982): 1007.
49. 15 U.S.C.A., sec. 522.
50. Johnson and Libecap, "Contracting Problems and Regulations," 1008.
51. De Alessi, "Fishing for Solutions," 109.
52. James L. Anderson and James E. Wilen, "Implications of Private Salmon Aquaculture on Prices, Production, and Management of Salmon Resources," *American Journal of Agricultural Economics* 68 (November 1986): 877.
53. Nelson Bryant, "A Scottish Group Protects Salmon," *New York Times,* January 8, 1990, S-13.
54. Terry L. Anderson and Donald R. Leal, *Enviro-Capitalists: Doing Good While Doing Well* (Lanham, MD: Rowman and Littlefield Publishers, 1997), 103.
55. Ibid., 135.
56. Sue Scott, "Greenland Salmon Fishery Ends," *News Release Communiqué,* Atlantic Salmon Federation, St. Andrews, New Brunswick, Canada, August 1, 1993.
57. Edwin S. Iversen and Jane Z. Iversen, "Salmon-farming Success in Norway," *Sea Frontiers* (November–October 1987): 355–61; Cheryl Sullivan, "Salmon Feedlots in Northwest," *Christian Science Monitor,* July 23, 1987. See also Robert R. Sticker, "Commercial Fishing and Net-pen Salmon Aquaculture: Turning Conceptual Antagonism Toward a Common Purpose," *Fisheries* 13 (July–August 1988): 9–13.
58. Carey Goldberg, "Fish Farms Breed Fight Over Way of Life," *New York Times,* website: http://www.nytimes.com, August 28, 1999; Margot Higgins, "Atlantic Salmon Protection Expected," *Environmental News Network,* October 15, 1999, website: http://www.enn.com/enn-news-archive/1999/10/101599/asalmon_6485.asp.
59. Letter and fact sheets on salmon breeding and habitat impact from Melissa Field, Schiedermayer & Associates, n.d.
60. Ibid.
61. Merrill Leffler, "Killing Maryland's Oysters," *Washington Post,* March 29, 1987.

62. Bruce Yandle, "The Commons: Tragedy or Triumph?" *The Freeman,* April 1999, 32.
63. "Artificial Reef 'Neptune' Complex Deployed," *Environment News Service,* May 26, 1999, on-line version, website: http://ens.lycos.com/e%2Dwire/may99/may269901.html.
64. De Alessi, *Fishing for Solutions,* 61.
65. Anthony D. Scott, "The ITQ as a Property Right: Where it Came From, How It Works, and Where It Is Going," in *Taking Ownership,* 97.
66. De Alessi, "Fishing for Solutions," 99.
67. Scott, "Market Solutions," 23.
68. Ronald N. Johnson, "Implications of Taxing Quota Value in an Individual Transferable Quota Fishery," *Marine Resource Economics* 10 (1995): 327-40.
69. De Alessi, "Fishing for Solutions," 108.
70. Gregory B. Christainsen and Brian C. Gothberg, "The Potential of High Technology for Establishing Tradeable Rights to Whales," paper presented at "The Technology of Property Rights," 1999 PERC Political Economy Forum, Bozeman, Montana, December 2–5, 1999.
71. Ibid.

CHAPTER 10

1. Lynn Scarlett, "Doing More with Less: Dematerialization—Unsung Environmental Triumph," in *Earth Report 2000,* ed. Ron Bailey (New York: McGraw-Hill, 2000), 54.
2. Lynn Scarlett and Jane S. Shaw, "Environmental Progress: What Every Executive Should Know," *PERC Policy Series* No. PS-15 (Bozeman, MT: Political Economy Research Center, April 1999), 3.
3. Lynn Scarlett, "New Environmentalism," *NCPA Policy Report* 201 (Dallas: National Center for Policy Analysis, January 1997), 11.
4. Indur Goklany, "Richer is Cleaner," in *The True State of the Planet,* ed. Ronald Bailey (New York: The Free Press, 1995), 348.
5. Iddo Wernick, Paul Waggoner, and Jesse Ausubel, "Searching for Leverage to Conserve Forests: The Industrial Ecology of Wood Products in the United States," *Journal of Industrial Ecology* 1(3): 125–45.
6. Livio D. DeSimone and Frank Popoff, *Eco-Efficiency: The Business Link to Sustainable Development* (Cambridge, MA: MIT Press, 1997), 2.
7. Ibid.
8. Jerald Blumber, Age Korsvold, and Georges Blum, "Environmental Performance and Shareholder Value," World Business Council for Sustainable Development, Geneva, Switzerland, 1996.
9. Bruce Van Voorst, "The Recycling Bottleneck," *Time,* September 14, 1992, 52–54. For more on "the recycling myth," see Michael Sanera and Jane S. Shaw, *Facts, Not Fear: Teaching Children About the Environment* (Washington, DC: Regnery Publishing, 1999), chapter 17.
10. Lynn Scarlett, Richard McCann, Robert Anex, and Alexander Volokh, "Packaging, Recycling, and Solid Waste," Reason Public Policy Institute, *Policy Study* 223, Los Angeles, 1997.
11. Quoted in Jim Carlton, "Going Green: Plastic Lumber Builds a Market," *Wall Street Journal,* March 31, 2000, B1.
12. See especially Julian Simon, *The Ultimate Resource* (Princeton, NJ: Princeton University Press, 1981).
13. Roger E. Meiners and Bruce Yandle, "The Common Law: How it Protects the Environment," *PERC Policy Series* No. PS-13 (Bozeman, MT: Political Economy Research Center, May 1998), 19.

14. See Indur Goklany, *Clearing the Air* (Washington, DC: Cato Institute, 1999), and Indur Goklany, "Empirical Evidence Regarding the Role of Nationalization in Improving U.S. Air Quality," in *The Common Law and the Environment,* ed. Roger E. Meiners and Andrew P. Morriss (Lanham, MD: Rowman and Littlefield Publishers, 2000).
15. Goklany, "Empirical Evidence," 48.
16. Ibid.
17. Richard L. Stroup, "Superfund: The Shortcut That Failed," in *Breaking the Environmental Policy Gridlock,* ed. Terry L. Anderson (Stanford, CA: Hoover Institution Press, 1997), 116.
18. U.S. General Accounting Office (GAO), *Superfund: Half the Sites Have All Cleanup Remedies in Place or Completed,* GAO/RCED-99–245, Washington, DC, July 1999, 5.
19. Stephen Breyer, *Breaking the Vicious Circle: Toward Effective Risk Regulation* (Cambridge, MA: Harvard University Press, 1993), 11.
20. W. Kip Viscusi and James T. Hamilton, "Are Risk Regulators Rational? Evidence from Hazardous Waste Cleanup Decisions," working paper no. 99–2, AEI-Brookings Joint Center for Regulatory Studies, Washington, DC, April 1999.
21. U.S. General Accounting Office (GAO), *Superfund: Progress, Problems and Future Outlook,* GAO/T-RCED-99–128, Washington, DC, March 23, 1999.
22. Bruce A. Ackerman and W. T. Hassler, *Clean Coal/Dirty Air, or How the Clean-Air Act Became a Multibillion-Dollar Bail-Out for High Sulfur Coal Producers and What Should Be Done About It* (New Haven: Yale University Press, 1981).
23. Robert W. Crandall, "Ackerman and Hassler's *Clean Coal/Dirty Air,*" *Bell Journal of Economics* 1 (autumn 1981): 678.
24. George Daly and Thomas Mayor, "Equity, Efficiency and Environmental Quality," *Public Choice* 51 (1986): 154.
25. For example, see Ackerman and Hassler, *Clean Coal/Dirty Air.*
26. Bruce Yandle, "Bootleggers, Baptists, and Global Warming," *PERC Policy Series* No. PS-14 (Bozeman, MT: Political Economy Research Center, November 1998), 6.
27. Ibid.
28. Nancie G. Marzulla and Roger J. Marzulla, *Property Rights: Understanding Government Takings and Environmental Regulation* (Rockville, MD: Government Institute, 1997), 93.
29. Bruce Yandle, "Environmental Regulation: Lessons from the Past and Future Prospects," in *Breaking the Environmental Policy Gridlock,* ed. Terry L. Anderson (Stanford, CA: Hoover Institution Press, 1997), 142.
30. Office of Management and Budget, *Analytical Perspectives, Budget of the United States Government, Fiscal Year 2000,* Washington, DC, 50.
31. David Schoenbrod, "Time for the Federal Environmental Aristocracy to Give Up Power," Center for the Study of American Business, *Policy Study* 144, Washington University, St. Louis, Missouri, February 1998, 8–9.
32. David Schoenbrod, "Protecting the Environment in the Spirit of the Common Law," in *The Common Law and the Environment: Rethinking the Statutory Basis for Modern Environmental Law,* ed. Roger E. Meiners and Andrew P. Morriss (Lanham, MD: Rowman and Littlefield Publishers, 2000), 3.
33. David Schoenbrod, "Why States, Not EPA, Should Set Pollution Standards," in *Environmental Federalism,* ed. Terry L. Anderson and Peter J. Hill (Lanham, MD: Rowman and Littlefield Publishers, 1997), 268.
34. BPT stands for "best practicable control technology available," as stated in the 1972 Clean Water Act.
35. Richard N. L. Andrews, *Managing the Environment, Managing Ourselves: A History of American Environmental Policy* (New Haven: Yale University Press, 1999), 270–71.
36. Ibid.
37. Ibid.

38. For a detailed discussion of this program, see David W. Riggs, "Market Incentive for Water Quality," in *The Market Meets the Environment,* ed. Bruce Yandle (Lanham, MD: Rowman and Littlefield Publishers, 1999), 167–204. Also see Sean Blacklocke, "Effluent Trading in South Carolina," in the same volume.
39. Riggs, "Market Incentive," 189.
40. James J. Opaluch and Richard M. Kashmanian, "Assessing the Viability of Marketable Permit Systems: An Application in Hazardous Waste Management," *Land Economics* 61 (August 1985): 263–71.
41. Ronald Coase, "The Problem of Social Cost," *Journal of Law and Economics* 3 (October 1960): 1–44.
42. For a complete discussion of the Love Canal story, see Eric Zuesse, "Love Canal: The Truth Seeps Out," *Reason* 12 (February 1981): 16–33, and Angela Ives, "Love Canal," in *The Market Meets the Environment,* 37–57.
43. Quoted in Ives, "Love Canal," 43.
44. Ibid., 48.
45. Fred L. Smith, Jr., "Controlling the Environmental Threat of the Global Liberal Order," paper presented to the Mont Pelerin Society, Christchurch, New Zealand, November 1989.
46. For a more complete discussion of contaminant source analysis for defining and enforcing property rights, see Anna M. Michalak, "Feasibility of Contaminant Source Identification for Property Rights Enforcement," paper presented at "The Technology of Property Rights," 1999 PERC Political Economy Forum, Bozeman, Montana, December 2–5, 1999.
47. See Mark Crawford, "Scientists Battle Over Grand Canyon Pollution," *Science* 247 (February 23, 1990): 911–12.
48. Interestingly, the EPA eventually showed that the Beatrice property was the most likely source of contamination, but it was Grace that paid compensation to the families.
49. 443 A.2d 1244 S.Ct., R.I. (1982).
50. Bruce Yandle, "Coase, Pigou, and Environmental Rights," in *Who Owns the Environment?* ed. Peter J. Hill and Roger E. Meiners (Lanham, MD: Rowman and Littlefield Publishers, 1998), 119–152.
51. Ibid., 138. Also see Bruce Yandle, *Common Sense and Common Law for the Environment* (Lanham, MD: Rowman and Littlefield Publishers, 1997).
52. See Andrew McFee Thompson, "Free Market Environmentalism and the Common Law: Confusion, Nostalgia, and Inconsistency," *Emory Law Journal* 45 (fall 1996): 1329–72.
53. Elizabeth Brubaker, "The Common Law and the Environment: The Canadian Experience," in *Who Owns the Environment?,* 92–93.
54. For a complete discussion of these cases, see Yandle, *Common Sense and Common Law.* Also see Roger E. Meiners and Bruce Yandle, "Common Law and the Conceit of Modern Environmental Policy," *George Mason Law Review* 7 (summer 1999): 923–82.
55. 94 F. 561, W.D. Ark., 1899.
56. 208 N.Y. 1, 101 N. E. 805 (1913).
57. 206 U.S. 230, 27 S.Ct. 618 (1907).
58. 265 F. 928 D. Utah (1919).
59. Yandle, *Common Sense and Common Law,* 108.
60. 406 U.S. 91 (1972).
61. 451 U.S. 304 (1981).
62. 459 N.Y.S.2d 971 (1983).
63. 456 N.Y.S.2d 867 (1982).

64. See Yandle, *Common Sense and Common Law,* 148.
65. Ibid., 146.
66. Meiners and Yandle, "The Common Law: How it Protects the Environment," 18–19.
67. For a discussion of these possibilities, see Murray Rothbard, "Law, Property Rights, and Air Pollution," *Cato Journal* 2 (spring 1982): 90.
68. Schoenbrod, "Protecting the Environment," 20.
69. Ibid., 21–22.

CHAPTER 11

1. Garrett Hardin, "The Tragedy of the Commons," *Science* 162 (December 1968): 1244.
2. Ibid.
3. The collapse of the Pacific sardine fishery provides a classic example. See J. L. McHugh, "Jeffersonian Democracy and the Fisheries," in *World Fisheries Policy: Multidisciplinary Views,* ed. B. J. Rothschild (Seattle: University of Washington Press, 1972), 134–55.
4. See chapter 3 in this volume.
5. Ibid.
6. Ibid., p. 34.
7. William Ophuls, "Leviathan or Oblivion?" in *Toward a Steady-State Economy,* ed. Herman Daley (San Francisco: W. H. Freeman, 1973), 228.
8. Richard J. Agnello and Lawrence P. Donnelley, "Property Rights and Efficiency in the Oyster Industry," *Journal of Law and Economics* 18 (1975): 521–33; Richard J. Agnello and Lawrence P. Donnelley, "Price and Property Rights in the Fisheries," *Southern Economic Journal* 42 (October 1979): 253–62.
9. For example, it would not make sense to divide up fishing grounds and assign property rights to the partitions, as fish are highly mobile and their concentrations unpredictable. Also, privatization can arouse intense opposition if it is seen as creating a "monopoly" to natural resource.
10. Svein Jentoft and Trond Kristoffersen, "Fishermen's Co-management: The Case of the Lofoten Fishery," *Human Organization* 48 (1989): 355.
11. Louis De Alessi, "Private Property Rights as the Basis for Free Market Environmentalism," in *Who Owns the Environment?* ed. Peter J. Hill and Roger E. Meiners (Lanham, MD: Rowman and Littlefield Publishers, 1998), 8.
12. Elinor Ostrom, *Governing the Commons: The Evolution of Institutions for Collective Action* (New York: Cambridge University Press, 1990); Terry L. Anderson and Randy T. Simmons, eds., *The Political Economy of Customs and Culture: Informal Solutions to the Commons Problem* (Lanham, MD: Rowman and Littlefield Publishers, 1993).
13. S. V. Ciriacy-Wantrup and Richard C. Bishop, "'Common Property' as a Concept in Natural Resources Policy," *Natural Resources Journal* 15 (1975): 13–27.
14. Robert Netting, *Balancing on an Alp* (New York: Cambridge University Press, 1981).
15. Margaret A. McKean, "Management of Traditional Common Lands (*Iriachi*) in Japan," in *Proceedings of the Conference on Common Property Resource Management, April 21–26, 1985,* National Research Council (Washington, DC: National Academy Press, 1986), 533–89.
16. Ostrom, *Governing the Commons,* 69–70.
17. Ibid., 35.
18. Ibid., 90.
19. Ibid., 93–94.
20. Ibid.

21. Elinor Ostrom, James Walker, and Roy Gardner, "Covenants With and Without a Sword: Self-Governance Is Possible," in *The Political Economy of Customs and Culture,* 127–156.
22. Ostrom, *Governing the Commons,* 101.
23. Marilyn B. Brewer, "Ingroup Bias in the Minimal Intergroup Situation: A Cognitive-Motivational Analysis," *Psychological Bulletin* 86 (1979): 307–24.
24. Many activities of young people require that parents and children participate in fund-raising activities, even though it might be more efficient for them to work at other independent jobs to raise money. A possible explanation for requiring participation is that it excludes those less interested in the activity and inculcates values that help overcome the free-rider problem.
25. For an explanation of this choice, see Terry L. Anderson and Fred McChesney, "Raid or Trade: An Economic Model of Indian-White Relations," *Journal of Law and Economics* 37 (April 1994): 39–74.
26. Sockeye salmon are found only in stream systems that include a freshwater lake.
27. R. L. Olson, "Social Structure and Social Life of the Tlingit in Alaska," *Anthropological Records* 26 (Berkeley: University of California Press, 1967).
28. Steve Langdon, "From Communal Property to Common Property of Limited Entry: Historical Ironies in the Management of Southeast Alaska Salmon," in *A Sea of Small Boats,* ed. John Cordell (Cambridge, MA: Cultural Survival, 1989), 304–32.
29. Kalervo Oberg, *The Social Economy of the Tlingit Indians,* American Ethnological Society Monograph 55 (Seattle: University of Washington Press, 1973), and Frederica De Laguna, *The Story of a Tlingit Community,* Bureau of American Ethnology Bulletin 172 (Washington, DC: U.S. Government Printing Office, 1972).
30. W. Goldsmith and T. H. Haas, "Possessory Rights of the Natives of Southeastern Alaska," report to the Commissioner of Indian Affairs, Bureau of Indian Affairs, U.S. Department of the Interior, Juneau, Alaska, 1946.
31. Langdon, "From Communal Property to Common Property of Limited Entry," 309.
32. In a 1986 journal article, D. Bruce Johnsen makes a compelling case that potlatching was instrumental in enforcing exclusive property rights to salmon streams. See D. Bruce Johnsen, "The Formation and Protection of Property Rights among Southern Kwakiutl Indians," *Journal of Legal Studies* 15 (January 1986): 41–67.
33. Harold Demsetz, "Toward a Theory of Property Rights," *American Economic Review* 57 (May 1967): 347–59 and Terry L. Anderson and Peter J. Hill, "The Evolution of Property Rights: A Study of the American West," *Journal of Law and Economics* 18 (1975): 163–79.
34. De Laguna, *The Story of a Tlingit Community,* 464.
35. Ibid.
36. Oberg, *The Social Economy,* 92–93.
37. Ibid., 63.
38. Goldsmith and Haas, "Possessory Rights of the Natives of Southeastern Alaska," 109, and Langdon, "From Communal Property to Common Property of Limited Entry," 314.
39. Langdon, "From Communal Property to Common Property of Limited Entry," 314.
40. Ibid., 318.
41. Robert Higgs, "Legally Induced Technical Regress in the Washington Salmon Fishery," *Research in Economic History* 7 (1982): 82.
42. Fikret Berkes, "Marine Inshore Fishery Management in Turkey," in *Proceedings of the Conference on Common Property Resource Management,* 63–83.
43. Ibid.
44. Ostrom, *Governing the Commons,*178-181.

45. Svein Jentoft, "Fisheries Co-management: Delegating Responsibility to Fishermen's Organizations," *Marine Policy* (April 1989): 142.
46. Cordell, ed., *A Sea of Small Boats,* 334.
47. William C. Herringbone, "Operation of the Japanese Management System," in *Alaska Fisheries Policy,* ed. Arleen R. Tusking, Thomas A. Morehouse, and James D. Babb, Jr. (Fairbanks: Institute of Social, Economic and Government Research, 1972), 421.
48. Kenneth Ruddle and Tomoya Akimichi, "Sea Tenure in Japan and the Southwestern Ryukus," in *A Sea of Small Boats,* 365.
49. H. Befu, "Political Ecology of Fishing in Japan: Techno-environmental Impact of Industrialization in the Inland Sea," *Research in Economic Anthropology* 3 (1980): 323–92.
50. Kevin MacEwen Short, "Self-management of Fishing Rights by Japanese Cooperative Associations: A Case Study from Hokkaido," in *A Sea of Small Boats,* 380.
51. Another case is the estuary fishery near Valensa, Brazil, discussed in chapter 9.
52. Francis P. Bowles and Margaret C. Bowles, "Holding the Line: Property Rights in the Lobster and Herring Fisheries of Matinicus Island, Maine," in *A Sea of Small Boats,* 229.
53. Ibid., 236.
54. Ibid., 243.
55. James Acheson, "Capturing the Commons: Legal and Illegal Strategies," in *The Political Economy of Customs and Culture,* 73.
56. Ibid., 74.
57. Ibid., 80.
58. Ibid.
59. Jerome B. Robinson, "The Next Step for Atlantic Salmon," *Field & Stream,* September 1994, 22–25.
60. Peter H. Pearse and James R. Wilson, "Local Co-management of Fish and Wildlife: The Quebec Experience," *Wildlife Society Bulletin* 27 (1999): 678.
61. Information provided by Yannick Routhier, Ministry of the Environment and Wildlife, Quebec, Canada, September 14, 1995.
62. Information provided by Yannick Routhier, November 22, 1994.
63. Jean-Francois Davignon, "The Quebec Story," Atlantic Salmon Federation, Quebec, Canada, n.d.
64. Pearse and Wilson, "Local Co-management of Fish and Wildlife," 683.
65. Even in developed countries such as the United States, a community-based wildlife management can take center stage. An exemplary case is the program carried out by the White Mountain Apache Tribe. See Terry L. Anderson and Donald R. Leal, *Enviro-Capitalists: Doing Good While Doing Well* (Lanham, MD: Rowman and Little field Publishers, 1997), 150–53.
66. Victoria Hylton, "The Wild Harvest," a document published by Southern Wild Productions, Johannesburg, South Africa, n.d., 2.
67. Ibid.
68. In Zimbabwe, a 1982 amendment to the 1975 Parks and Wildlife Act spelled out that district councils could be designated as "appropriate authorities" for managing wildlife. See CAMPFIRE (Fact Sheets), "Acts, Amendments & Appropriate Authorities: Campfire's legal framework," March 24, 2000, website: http://www.campfire-zimbabwe.org/facts_03.html
69. Conversions of Zimbabwean dollars were made assuming Z$1 is equivalent to U.S.$0.40.
70. Valerie Thresher, "Economic Reflections on Wildlife Utilization in Zimbabwe," master's thesis, University of California at Davis, 1993, 45.
71. Ibid., 50.

72. Environmental Consultants (Pvt) Ltd., *People, Wildlife and Natural Resources—The CAMPFIRE Approach to Rural Development in Zimbabwe* (Harare, Zimbabwe: Zimbabwe Trust, 1990), 23.
73. Gregory F. Maggio, "Recognizing the Vital Role of Local Communities in International Legal Instruments for Conserving Biodiversity," *UCLA Journal of Environmental Law & Policy* 16 (1997/1998): 199.
74. The Convention on International Trade in Endangered Species of Wild Fauna and Flora (CITES) became effective on July 1, 1975, and has steadily grown to 130 signed parties.
75. At the Tenth Conference of Parties of CITES, held in Zimbabwe in June of 1997, the strict ban on the elephant-ivory trade was lifted. Over objections from animal rights groups, empirical evidence indicating that limited harvests of elephant ivory would increase the health of both elephants and local community economies convinced convention parties to move the African elephant to Appendix II categorization, which allows regulated trade in elephant ivory. See Sean T. McAllister, "Community-Based Conservation: Restructuring Institutions to Involve Local Communities in a Meaningful Way," *Colorado Journal of International Environmental Law & Policy* 10 (winter 1999): 220.
76. Ostrom, *Governing the Commons,* 65.
77. Jared Diamond, "Paradise and Oil: In a New Guinea Rain Forest, Environmentalists and Business Executives Learn that What's Good for This Pristine World is Also Good for the Bottom Line," *Discover,* March 1999, 96.
78. Peter Eaton, "Customary Land Tenure and Conservation in Papua New Guinea," in *Culture and Conservation: The Human Dimension in Environmental Planning,* ed. Jeffery A. McNeely and David Pitt (Dover, NH: Croom Helm, 1985), 181–191.
79. Ibid., 101.
80. Chevron Corporation, "Respecting Rivers and Rain Forests," website: http://www.chevron.com/environment/index.html.
81. John Stackhouse, "Forests Returning to the Himalayas: First Nepal's Forestry Program Failed, Then the People Took Over and Saved the Trees," *Globe and Mail,* October 22, 1998.
82. D. A. Messerschmidt, "Collective Management of Forest Hills in Nepal," in *Proceedings of the Conference on Common Property Resource Management,* 458.
83. Stackhouse, "Forests Returning to the Himalayas."
84. Ibid.
85. Robert B. Keiter, "Preserving Nepal's National Parks: Law and Conservation in the Developing World," *Ecology Law Quarterly* 22 (1995): 637.
86. Ibid., 648.

CHAPTER 12

1. For a general discussion of the globalization of environmental issues, see Terry L. Anderson and Henry I. Miller, eds., *The Greening of U.S. Foreign Policy* (Stanford, CA: Hoover Institution Press, 2000).
2. For example, see Robert Balling, Jr., *The Heated Debate: Greenhouse Predictions Versus Climate Reality* (San Francisco: Pacific Research Institute for Public Policy, 1992), and Roy W. Spencer, "How Do We Know the Temperature of the Earth?" in *Earth Report 2000: Revisiting the True State of the Planet,* ed. Ronald Bailey (New York: McGraw-Hill, 2000), 23–40.
3. Wayne A. Morrisey and John R. Justus, "IB89005: Global Climate Change," CRS Issue Brief for Congress, Congressional Research Service, March 13, 2000, on-line version, website: http://www.cnie.org/nle/clm-2.html.

4. Spencer, "How Do We Know The Temperature," 24.
5. For a discussion of the potential costs of the Kyoto Protocol, see Bruce Yandle, "Bootleggers, Baptists, and Global Warming," *PERC Policy Series* No. PS-14 (Bozeman, MT: Political Economy Research Center, November 1998).
6. Thomas Gale Moore, *Climate of Fear* (Washington, DC: Cato Institute, 1998), 142–45.
7. Sandra S. Batie, "Sustainable Development: Challenges to the Profession of Agricultural Economics," *American Journal of Agricultural Economics* 71 (December 1989): 1084–1101. See also a new journal called *Ecological Economics,* published by the International Society for Ecological Economics, c/o Burk and Associates, Inc., 1313 Dolley Madison Blvd., Suite 402, McLean VA 22101.
8. Batie, "Sustainable Development," 1085.
9. Kenneth Boulding, "The Economics of the Coming Spaceship Earth," in *Environmental Quality in a Growing Economy,* ed. H. Jarret (Baltimore: Johns Hopkins University Press, for Resources for the Future, 1966), 3–14; Herman E. Daly, *Steady-State Economics* (San Francisco: W. H. Freeman and Company, 1977).
10. Anderson and Miller, *The Greening of U.S. Foreign Policy.*
11. Timothy J. O'Riordan, "The Politics of Sustainability," in *Sustainable Environmental Management,* ed. R. K. Rutner (Boulder, CO: Westview Press, 1988).
12. Batie, "Sustainable Development," 1084, 1085.
13. Harold Barnett and Chandler Morse, *Scarcity and Growth: The Economics of Natural Resource Availability* (Baltimore: Johns Hopkins University Press, for Resources for the Future, 1963), 249.
14. Lynn Scarlett and Jane S. Shaw, "Environmental Progress: What Every Executive Should Know," *PERC Policy Series* No. PS-15 (Bozeman, MT: Political Economy Research Center, April 1999), 17–19.
15. Tom Tietenberg, *Environmental and Natural Resource Economics* (Glenview, IL: Scott, Foresman and Company, 1984), 437.
16. Gro Harlem Burndtland, "From the Cold War to a Warm Atmosphere," *New Perspectives Quarterly* 6 (1989): 5.
17. H. Crane Miller, *Turning the Tide on Wasted Tax Dollars: Potential Federal Savings from Additions to the Coastal Barrier Resources System* (Washington, DC: National Wildlife Federation, April 17, 1989).
18. Robert Repetto, *The Forest for the Trees? Government Policies and the Misuse of Forest Resources* (Washington, DC: World Resources Institute, 1988), 17–32. See also "How Brazil Subsidizes the Destruction of the Amazon," *The Economist,* March 18, 1989, 69.
19. John O. Browder, "Public Policy and Deforestation in the Brazilian Amazon," in *Public Policies and the Misuse of Forest Resources,* ed. Robert Repetto and Malcolm Gillis (Cambridge, England: Cambridge University Press, 1988), 251–52.
20. Robert T. Deacon, "Deforestation and the Rule of Law in a Cross Section of Countries," *Land Economics* 70 (1994): 414–30.
21. Lee J. Alston, Gary D. Libecap, and Bernardo Mueller, *Titles, Conflict, and Land Use: The Development of Property Rights and Land Reform on the Brazilian Amazon Frontier* (Ann Arbor: University of Michigan Press, 1999).
22. Gareth Porter, "Too Much Fishing Fleet, Too Few Fish: A Proposal for Eliminating Global Fishing Overcapacity," a prepublication draft, World Wildlife Fund, August 1998, 22.
23. Food and Agriculture Organization of the United Nations (FAO). *Marine Fisheries and the Law of the Sea: A Decade of Change* (Rome, Italy: FAO, 1993), 32.
24. Mateo Milazzo, "Subsidies in World Fisheries: A Reexamination," World Bank Technical Paper No. 406 (Fisheries Series) (Washington, DC: National Academy Press, 1999), 74.

25. James M. Sheehan, "The Greening of the World Bank: A Lesson in Bureaucratic Survival," Foreign Policy Briefing No. 56 (Washington, DC: Cato Institute, 2000).
26. Bruce Rich, *Mortgaging the Earth* (Boston: Beacon Press, 1994).
27. James Gwartney, Robert Lawson, and Walter Block, *Economic Freedom of the World, 1975–1995* (Vancouver, BC, Canada: The Fraser Institute, 1996), 93–94.
28. William W. Beach and Gareth Davis, "The Index of Economic Freedom and Economic Growth," in *1997 Index of Economic Freedom,* ed. Kim R. Holmes, Bryan T. Johnson, and Melanie Kirkpatrick (Washington, DC: Heritage Foundation, 1997), 9.
29. William C. Clark, "Witches, Floods, and Wonder Drugs: Historical Perspectives on Risk Management," in *Societal Risk Assessment: How Safe Is Safe Enough?* ed. Richard C. Schwing and Walter A. Albers, Jr. (New York: Plenum Press, 1988), chapter 4; Aaron Wildavsky, *Searching for Safety* (New Brunswick, NJ: Transaction Books, 1988), chapter 4.
30. Wildavsky, *Searching for Safety,* chapter 3.
31. Seth W. Norton, "Property Rights, the Environment, and Economic Well-Being," in *Who Owns the Environment?* ed. Peter J. Hill and Roger E. Meiners (Lanham, MD: Rowman and Littlefield Publishers, 1998), 37-54.
32. Ibid., 51.
33. For a complete discussion, see Terry L. Anderson and Randy T. Simmons, eds., *The Political Economy of Customs and Culture: Informal Solutions to the Commons Problem* (Lanham, MD: Rowman and Littlefield Publishers, 1993).
34. Julian Morris, "International Environmental Agreements: Developing Another Path," in *The Greening of U.S. Foreign Policy.*
35. 3 RIIA 1905 (1949), 1965.
36. "The Mess One Man Makes," *The Economist,* April 22, 2000, 19–22.
37. For abundant examples of how modern technology is facilitating the evolution of property rights, see Terry L. Anderson and Peter J. Hill, eds., *The Technology of Property Rights* (Lanham, MD: Rowman and Littlefield Publishers, forthcoming).
38. George Francis, "Great Lakes Governance and the Ecosystem Approach: Where Next?" *Alternatives* 3 (September–October 1986): 66.
39. "Green Economics," *The Economist,* June 24, 1989, 48.

CHAPTER 13

1. Mikhail S. Bernstam, *The Wealth of Nations and the Environment* (London, England: Institute of Economic Affairs, 1991), 22.
2. Ibid., 24.
3. Gareth W. Dodd, "EPA Report Says U.S. Government Nation's No. 1 Polluter of Waterways," *U.S. Water News* (May 2000): 12.
4. Holly Lippke Fretwell, *Public Lands: The Price We Pay,* Public Lands Report No. 1 (Bozeman, MT: Political Economy Research Center, August 1998); and *Forests: Do We Get What We Pay For?* Public Lands Report No. 2 (Bozeman, MT: Political Economy Research Center, July 1999).
5. Holly Lippke Fretwell, *Federal Estate: Is Bigger Better?* Public Lands Report No. 3 (Bozeman, MT: Political Economy Research Center, May 2000), 2.
6. For details on these and other examples, see chapter 5 in this volume.
7. Traci Watson, "Environmental Groups Wielding Power of the Purse," *USA Today,* February 3, 2000.
8. This term is from the title of Hank Fischer's book, *Wolf Wars: The Remarkable Story of the Restoration of Wolves in Yellowstone* (Helena, MT: Falcon Publishing, 1995).
9. For details, see Terry L. Anderson, "Home on the Range for Wolves," *Christian Science Monitor,* April 14, 1994.

10. Defenders of Wildlife, "Wolf Compensation Trust," May 12, 2000, website: http://www.defenders.org/wolfcomp.html.
11. Quoted in Watson, "Environmental Groups Wielding Power of the Purse."
12. Anderson, "Home on the Range for Wolves."
13. Delta Waterfowl, *Delta Waterfowl Report,* Deerfield, Illinois, fall 1994.
14. Karol Jablonski, Delta Waterfowl Foundation, telefax, May 15, 2000.
15. Delta Waterfowl, *Delta Waterfowl: Adopt a Pothole Summary Report,* Deerfield, Illinois, August 1993.
16. Donald R. Leal, "Unlocking the Logjam Over Jobs and Endangered Animals," *San Diego Union-Tribune,* April 18, 1993.
17. For a discussion of the difference between federal and state land management, see Donald R. Leal, "Turning a Profit on Public Forests," *PERC Policy Series* No. PS-4 (Bozeman, MT: Political Economy Research Center, September 1995).
18. Political Economy Research Center, *Commentary by Northwest Ecosystem Alliance,* July 7, 1999, website: http://www.perc.org/newsloom.htm.
19. Matthew Brown and Jane S. Shaw, "Viewpoint: Paying to Prevent Logging is Breakthrough for Environmentalists," *Tribnet,* August 13, 1998, website: http://tribnet.com/nesw/oped/0813a111.
20. Quoted in Brown and Shaw, "Viewpoint: Paying to Prevent Logging."
21. Linda Platts, "Environmentalists Use Market Tools," *Montana Farmer-Stockman,* February 1997, 18.
22. Lisa Church, "Fun Hogs to Replace Cows in a Utah Monument," *High Country News,* February 1, 1999, 4.
23. Quoted in Church, "Fun Hogs to Replace Cows in Utah Monument."
24. Platts, "Environmentalists Use Market Tools."
25. Watson, "Environmental Groups Wielding Power of the Purse."
26. For a complete discussion of water regulations, see chapter 8 in this volume.
27. See chapter 8 in this volume.
28. See chapter 8 in this volume.
29. Quoted in David Stalling, "Public Elk, Private Lands: Should Landowners Benefit from Elk and Elk Hunting?" *Bugle,* January–February 1999, 73.
30. Dennis Glick, David Cowan, Robert Bonnie, David Wilcove, Chris Williams, Dominick Dellasala, and Steve Primm, *Incentives for Conserving Open Lands in Greater Yellowstone* (Bozeman, MT: Greater Yellowstone, 1998), 8.
31. Ibid., 19.
32. Aldo Leopold, "Conservation Economics," in *The River of the Mother of the Mother of God and Other Essays by Aldo Leopold,* ed. Susan L. Flader and J. Baird Callicott (Madison: University of Wisconsin Press, 1991 [1934]), 202.
33. Terry L. Anderson, Vernon L. Smith, and Emily Simmons, "How and Why to Privatize Federal Lands," *Cato Policy Analysis* 363, December 9, 1999.
34. Donald R. Leal and Holly Lippke Fretwell, "Back to the Future to Save Our Parks," *PERC Policy Series* No. PS-10 (Bozeman, MT: Political Economy Research Center, June 1997), 9.
35. Donald R. Leal and Holly Lippke Fretwell, "Parks in Transition: A Look at State Parks," Bozeman, MT: Political Economy Research Center, June 1997, website: http://www/perc.org/stpk.htm.
36. John Baden and Richard Stroup, "Saving the Wilderness: A Radical Proposal," *Reason* 13 (July 1981): 28-36.
37. Presidio Trust, *The Presidio Trust Financial Management Program,* report to Congress, San Francisco, California, July 8, 1998, 3.
38. As of May 2000, legislation is pending in Congress to acquire and apply a trust

approach to the 95,000–acre Baca Ranch in New Mexico. Generally, when the federal government acquires land, it can spell bad news given its track record of mismanaging resources under its care. In this case, however, an experimental management regime is being provided that is intended to be cost effective and environmentally sensitive.

39. See chapter 10 in this volume.
40. U.S. Environmental Protection Agency, *Draft Framework for Watershed-Based Trading,* EPA 800–R-96–001, Washington, DC: U.S. Environmental Protection Agency, May 1996.
41. Bruce Yandle, "Environmental Regulation: Lessons from the Past and Future Prospects," in *Breaking the Environmental Gridlock,* ed. Terry L. Anderson (Stanford, CA: Hoover Institution Press, 1997), 161–62.
42. Sand County Foundation, telefax, July 28, 1995.
43. Quoted in Watson, "Environmental Groups Wielding the Power of the Purse."
44. Terry L. Anderson and Donald R. Leal, *Enviro-Capitalists: Doing Good While Doing Well* (Lanham, MD: Rowman and Littlefield Publishers, 1997), 48–52.
45. Ibid., 79–82.
46. Terry L. Anderson and Pamela S. Snyder, *Water Markets: Priming the Invisible Pump* (Washington, DC: Cato Institute, 1997).
47. Anderson and Leal, *Enviro-Capitalists,* 69–73.
48. Terry L. Anderson and Alexander James, eds., *The Politics and Economics of Park Management* (Lanham, MD: Rowman and Littlefield Publishers, 2001).
49. Peter S. Menell, "Institutional Fantasylands: From Scientific Management to Free Market Environmentalism," *Harvard Journal of Law and Public Policy* 15 (1992): 589–610.
50. David Schoenbrod, "Protecting the Environment in the Spirit of the Common Law," in *The Common Law and the Environment: Rethinking the Statutory Basis for Modern Environmental Law,* ed. Roger E. Meiners and Andrew P. Morriss (Lanham, MD: Rowman and Littlefield Publishers, 2000), 3–24.

BIBLIOGRAPHY

Abbott, Jr., Gordon. "Long-Term Management: Problems and Opportunities." In *Private Options: Tools and Concepts for Land Conservation,* edited by Barbara Rusmore, Alexandra Swaney, and Allan D. Spader. Covello, CA: Island Press, 1982.

Acheson, James J. "Capturing the Commons: Legal and Illegal Strategies." In *The Political Economy of Customs and Culture: Informal Solutions to the Commons Problem,* edited by Terry L. Anderson and Randy T. Simmons. Lanham, MD: Rowman and Littlefield Publishers, 1993.

Ackerman, Bruce A., and W. T. Hasler, *Clean Coal/Dirty Air, or How the Clean-Air Act Became a Multibillion-Dollar Bail-Out for High Sulfur Coal Producers and What Should Be Done About It.* New Haven: Yale University Press, 1981.

"Acts, Amendments & Appropriate Authorities: Campfire's legal framework." CAMPFIRE (Fact Sheets). Available: http://www.campfire-zimbabwe.org/facts_03.html. Accessed: March 24, 2000.

Agnello, Richard J., and Lawrence P. Donnelley. "Property Rights and Efficiency in the Oyster Industry." *Journal of Law and Economics* 18 (1975): 521–33.

———. "Price and Property Rights in the Fisheries." *Southern Economic Journal* 42 (October 1979): 253–62.

Allison, Lee. State Geologist, Utah Geological Survey. Telephone conversation. August 19, 1997.

Alston, Lee J., Gary D. Libecap, and Bernardo Mueller. *Titles, Conflicts, and Land Use: The Development of Property Rights and Land Reform on the Brazilian Amazon Frontier.* Ann Arbor: University of Michigan Press, 1999.

American Forestry Association. *Proceedings of the American Forest Congress.* Washington, DC, 1905.

American Petroleum Institute. *Should Federal Onshore Oil and Gas Be Put Off Limits?* Washington, DC: American Petroleum Institute, June 1984.

American Petroleum Institute. "Access to Government Lands For Increased U.S. Oil Development." July 14, 1998. Available: http://www.api.org/news/backup/496access.htm.

Anderson, James L., and James E. Wilen. "Implications of Private Salmon Aquaculture on Prices, Production, and Management of Salmon Resources." *American Journal of Agricultural Economics* 68 (November 1986): 877.

Anderson, Terry L., and Alexander James, eds. *The Politics and Economics of Park Management.* Lanham, MD: Rowman and Littlefield Publishers, 2001.

Anderson, Terry L., and Donald R. Leal. "A Private Fix for Leaky Trout Streams." *Fly Fisherman* 19 (June 1988): 28–31.

———. *Enviro-Capitalists: Doing Good While Doing Well.* Lanham, MD: Rowman and Littlefield Publishers, 1997.

Anderson, Terry L., and Fred McChesney. "Raid or Trade: An Economic Model of Indian-White Relations." *Journal of Law and Economics* 37 (April 1994): 39–74.

Anderson, Terry L., and Henry I. Miller, eds. *The Greening of U.S. Foreign Policy.* Stanford, CA: Hoover Institution Press, 2000.

Anderson, Terry L., and Holly Lippke Fretwell. "A Trust for Grand Staircase-Escalante." *PERC Policy Series* No. PS-16, Bozeman, MT: Political Economy Research Center, September 1999.

Anderson, Terry L., and Jane Shaw. "Grass Isn't Always Greener in a Public Park." *Wall Street Journal,* May 28, 1985, 30.

Anderson, Terry L., and P. J. Hill. "Privatizing the Commons: An Improvement?" *Southern Economic Journal* 50 (1983): 438–50.

———. "The Race for Property Rights." *Journal of Law and Economics* 33 (April 1990): 177–97.

Anderson, Terry L., and Pamela S. Snyder. "A Free Market Solution to Groundwater Allocation in Georgia." *Issue Analysis.* Atlanta: Georgia Public Policy Foundation, 1996.

———. *Water Markets: Priming the Invisible Pump.* Washington, DC: Cato Institute, 1997.

———. "Priming the Invisible Pump." *PERC Policy Series* No. PS-9. Bozeman, MT: Political Economy Research Center, February 1997.

Anderson, Terry L., and Peter J. Hill. "The Evolution of Property Rights: A Study of the American West." *Journal of Law and Economics* 18 (1975): 163–79.

———. "From Free Grass to Fences: Transforming the Commons of the American West." In *Managing the Commons,* edited by Garrett Hardin and John Baden. San Francisco: W. H. Freeman, 1977.

———. *The Birth of a Transfer Society.* Stanford, CA: Hoover Institution Press, 1980.

———. "Privatizing the Commons: An Improvement?" *Southern Economics Journal* 50 (October 1983): 438–50.

———, eds. *The Technology of Property Rights.* Lanham, MD: Rowman and Littlefield Publishers, forthcoming.

Anderson, Terry L., and Randy T. Simmons, eds. *The Political Economy of Customs and Culture: Informal Solutions to the Commons Problem.* Lanham, MD: Rowman and Littlefield Publishers, 1993.

Anderson, Terry L., and Ronald N. Johnson. "The Problem of Instream Flows." *Economic Inquiry* 24 (October 1986): 535–54.

Anderson, Terry L., Oscar Burt, and David Fractor. "Privatizing Groundwater Basins: A Model and Its Applications." In *Water Rights: Scarce Resource Allocation, Bureaucracy, and the Environment,* edited by Terry L. Anderson. San Francisco: Pacific Institute for Public Policy Research, 1983.

Anderson, Terry L., Vernon L. Smith, and Emily Simmons. "How and Why to Privatize Federal Lands." *Cato Policy Analysis* 363, December 9, 1999.

Anderson, Terry L. "Institutional Underpinnings of the Water Crisis." *Cato Journal* 2 (winter 1983): 759–92.

———. "Camped Out in Another Era." *Wall Street Journal,* January 14, 1987.

———. "Home on the Range for Wolves." *Christian Science Monitor,* April 14, 1994.

———. "To Fee or Not to Fee: The Economics of Below-Cost Recreation." In *Multiple Conflicts Over Multiple Uses,* edited by Terry L. Anderson. Bozeman, MT: Political Economy Research Center, 1994.

———. "Water Options for the Blue Planet." In *The True State of the Planet,* edited by Ronald Bailey. New York: Free Press, 1995.

———. "Conservation—Native American Style." *PERC Policy Series* No. PS-6. Bozeman, MT: Political Economy Research Center, July 1996.

Andrews, Richard N. L. *Managing the Environment, Managing Ourselves: A History of American Environmental Policy.* New Haven: Yale University Press, 1999.

Arnason, Ragnar. "Property Rights as an Organizational Framework in Fisheries: The Cases of Six Fishing Nations." In *Taking Ownership: Property Rights and Fishery Management on the Atlantic Coast,* edited by Brian Lee Crowley. Halifax, Nova Scotia: Atlantic Institute for Market Studies, 1996.

Arno, Steve. "The Concept: Restoring Ecological Structure and Process in Ponderosa Pine Forests." In *The Use of Fire in Forest Restoration.* USFS Intermountain Research Station INT-6TR-341, June 1996.

"Artificial Reef 'Neptune' Complex Deployed." *Environment News Service,* May 26, 1999. Available: http://ens.lycos.com/e%2Dwire/may99/may269901.html.

"As the Cattle Business Weakens, Ranchers Turn Their Land Over to Recreational Use." *Wall Street Journal,* August 27, 1985, 33.

Ashe, Victor H. "Needs and Opportunities for Outdoor Recreation." In *Transactions of the Fifty-first North American Wildlife and Natural Resources Conference,* edited by Richard E. McCabe. Washington, DC: Wildlife Management Institute, 1986.

"Babbitt Endorses Water Markets and Places High Premium on Political Consensus." *Water Intelligence Monthly* (January 1996): 2–3.

Baden, John, and Richard Stroup. "Saving the Wilderness: A Radical Proposal." *Reason* 13 (July 1981): 28–36.

Baen, John S. *The Growing Importance and Value Implications of Recreational Hunting Leases to Agricultural Land Investors in America.* Presentation at American Real Estate Society meeting, Sarasota, Florida, April 16–19, 1997. Available from John S. Baen, Associate Professor of Real Estate, College of Business Administration, Finance, Insurance, Real Estate and Law, University of North Texas, Denton, Texas.

Bailey, Ronald. "Earth Day, Then and Now." *Reason On Line.* Available: http://www.reson.com/0005/fe.rb.earth.html.

Bailey, Ronald, ed. *Earth Report 2000: Revisiting the True State of the Planet.* New York: McGraw-Hill, 2000.

Balling, Jr., Robert. *The Heated Debate: Greenhouse Predictions Versus Climate Reality.* San Francisco: Pacific Research Institute for Public Policy, 1992.

Barnett, Harold, and Chandler Morse. *Scarcity and Growth: The Economics of Natural Resource Availability.* Baltimore: Johns Hopkins University Press, for Resources for the Future, 1963.

Baskin, Ken. "The Tug of War for the Wilderness." *Sun* (autumn 1985): 7.

Batie, Sandra S. "Sustainable Development: Challenges to the Profession of Agricultural Economics." *American Journal of Agricultural Economics* 71 (December 1989): 1084-1101.

Bator, Francis. "The Simple Analytics of Welfare Maximization." *American Economic Review* (March 1957): 22–59.

Beach, William W., and Gareth Davis. "The Index of Economic Freedom and Economic Growth." In *1997 Index of Economic Freedom,* edited by Kim R. Holmes, Bryan T. Johnson, and Melanie Kirkpatrick. Washington, DC: Heritage Foundation, 1997.

Befu, H. "Political Ecology of Fishing in Japan: Techno-environmental Impact of Industrialization in the Inland Sea." *Research in Economic Anthropology* 3 (1980): 323–92.

Bell, Frederick W. "Technological Externalities and Common-Property Resources: An Empirical Study of the U.S. Northern Lobster Fishery." *Journal of Political Economy* 80 (January–February 1972): 148–58.

Benjamin, Daniel K., Kurtis J. Swope, and Terry L. Anderson. "Bucks for Ducks or Money for Nothin'? The Political Economy of the Federal Duck Stamp Program." In *Political Environmentalism: Going Behind the Green Curtain,* edited by Terry L. Anderson. Stanford, CA: Hoover Institution Press, 2000.

Berkes, Fikret. "Marine Inshore Fishery Management in Turkey." In *Proceedings of the Conference on Common Property Resource Management.* National Research Council. Washington, DC: National Academy Press, 1986.

Bernstam, Mikhail S. *The Wealth of Nations and the Environment.* London, England: Institute of Economic Affairs, 1991.

———"Comparative Trends in Resource Use and Pollution in Market and Socialist Economies." In *The State of Humanity,* edited by Julian Simon. Cambridge, MA: Blackwell Publishers, 1995.

Bishop, C. A., E. F. Murphy, M. B. Davis, J. W. Baird, and G. A. Rose. *An Assessment of the Cod Stock in NAFO Divisions 2j+3k.* NAFO Scientific Council Research Document 93/86, serial number N2771, 1993.

Blacklocke, Sean. "Effluent Trading in South Carolina." In *The Market Meets the Environment,* edited by Bruce Yandle. Lanham, MD: Rowman and Littlefield Publishers, 1999.

Blood, Tom, and John Baden. "Wildlife Habitat and Economic Institutions: Feast or Famine for Hunters and Game." *Western Wildlands* 10 (spring 1984): 13.

Blumber, Jerald, Age Korsvold, and Georges Blum. "Environmental Performance and Shareholder Value." World Business Council for Sustainable Development, Geneva, Switzerland, 1996.

Boitnott, Richard A. Manager, Wildlife Ecology, International Paper Company, Mid-South Region, Shreveport, Louisiana. Letter to author. March 8, 1994.

———. International Paper Company. Personal communication, 1999.

Boulding, Kenneth. "The Economics of the Coming Spaceship Earth." In *Environmental Quality in a Growing Economy,* edited by H. Jarret. Baltimore: Johns Hopkins University Press, for Resources for the Future, 1966.

Bourland, Tom, Crawford and Bourland Consulting Foresters. Personal communication, 1999.

Bowles, Francis P., and Margaret C. Bowles. "Holding the Line: Property Rights in the Lobster and Herring Fisheries of Matinicus Island, Maine." In *A Sea of Small Boats,* edited by John Cordell. Cambridge, MA: Cultural Survival, 1989.

Brewer, Marilyn B. "Ingroup Bias in the Minimal Intergroup Situation: A Cognitive-Motivational Analysis." *Psychological Bulletin* 86 (1979): 307–24.

Breyer, Stephen. *Breaking the Vicious Circle: Toward Effective Risk Regulation.* Cambridge, MA: Harvard University Press, 1993.

Bromley, Daniel W. *Property Rights and the Environment: Natural Resource Policy in Transition.* Cambridge, MA: Blackwell, 1991.

Browder, John O. "Public Policy and Deforestation in the Brazilian Amazon." In *Public Policies and the Misuse of Forest Resources,* edited by Robert Repetto and Malcolm Gillis. Cambridge, England: Cambridge University Press, 1988.

Brown, Lester R. "The Future of Growth." In *State of the World 1998.* Project directed by Lester R. Brown. New York: W. W. Norton & Company, 1998.

Brown, Matthew, and Jane S. Shaw. "Viewpoint: Paying to Prevent Logging is Breakthrough for Environmentalists." *Tribnet,* August 13, 1998. Available: http://tribnet.com/nesw/oped/0813a111.

Brubaker, Elizabeth. "The Common Law and the Environment: The Canadian Experience." In *Who Owns the Environment?* edited by Peter J. Hill and Roger E. Meiners. Lanham, MD: Rowman and Littlefield Publishers, 1998.

———. "Property Rights: Creating Incentives and Tools for Sustainable Fisheries Management." *Fraser Forum.* BC, Canada: The Fraser Institute (April 1998): 10.

Bryant, Nelson. "Fishing Licenses Are At Issue." *New York Times,* February 5, 1989.

———. "A Scottish Group Protects Salmon." *New York Times,* January 8, 1990, S-13.

Bureau of the Census. *The Statistical History of the United States: From Colonial Times to the Present.* New York: Basic Books, 1976.

Burndtland, Gro Harlem. "From the Cold War to a Warm Atmosphere." *New Perspectives Quarterly* 6 (1989): 5.

Burt, Oscar, and Ronald G. Cummings. "Production and Investment in Natural Resource Industries." *American Economic Review* 60 (1970): 576–90.

"California Oil Spill Endangers Wildlife: Headway Made in Cleanup." CNN, Los Angeles. September 30, 1997. Available: http://www.cnn.com/US/9709/30/calif.oil.spill/.

Carlton, Jim. "Going Green: Plastic Lumber Builds a Market." *Wall Street Journal,* March 31, 2000, B1.

Causey, Don, ed. "Found! A Season-Long Deer Hunt For Only $700!" *U.S. Hunting Report* (June 1997): 1.

Causey, Don. "This State's Landowner Program A Real Winner." *U.S. Hunting Report* (June 1998): 8–10.

Center for Responsive Politics. "Lobbyists Spending in Washington." Available: http://www.opensecrets.org/lobbyists/98lookup.htm. Accessed: March 10, 2000.

Chandler, William J., ed. *Audubon Wildlife Report, 1988/1989.* San Diego: Academic Press, 1988.

Chase, Alston. *Playing God in Yellowstone.* Boston: Atlantic Monthly Press, 1986.

Chevron Corporation. "Respecting Rivers and Rain Forests." Available: http://www.chevron.com/environment.

Christainsen, Gregory B., and Brian C. Gothberg. "The Potential of High Technology for Establishing Tradeable Rights to Whales." Paper presented at "The Technology of Property Rights," 1999 PERC Political Economy Forum, Bozeman, Montana, December 2–5, 1999.

Christy, Jr., Francis T. "Paradigm Lost: The Death Rattle of Open Access and the Advent of Property Rights to Regimes in Fisheries." Paper prepared for the Eighth Biennial Conference of the Institute of Fisheries Economics and Trade, Marrakesh, Morocco, July 1–4, 1996.

Christy, Jr., Francis T., and Anthony Scott. *The Common Wealth in Ocean Fisheries.* Baltimore: Johns Hopkins University Press, for Resources for the Future, 1965.

Church, Lisa. "Fun Hogs to Replace Cows in a Utah Monument." *High Country News,* February 1, 1999, 4.

Ciriacy-Wantrup, S. V., and Richard C. Bishop. "Common Property as a Concept in Natural Resources Policy." *Natural Resources Journal* 15 (1975): 13–27.

Clark, Colin W. "Profit Maximization and the Extinction of Animal Species." *Journal of Political Economy* 81 (August 1981): 950–60.

Clark, William C. "Witches, Floods, and Wonder Drugs: Historical Perspectives on Risk Management." In *Societal Risk Assessment: How Safe Is Safe Enough?* edited by Richard C. Schwing and Walter A. Albers, Jr. New York: Plenum Press, 1988.

Clarke, Brian. "The Nymph in Still Water." In *The Masters of the Nymph,* edited by J. M. Migel and L. M. Wright. New York: Nick Lyons, 1979.

Clawson, Marion. *The Federal Lands Revisited.* Washington, DC: Resources for the Future, 1983.

"Clinton Extends Moratorium On Offshore Oil Drilling." CNN. June 1998. Available: http://www-cgi.cnn.com/TECH/science/9806/12/offshore.drilling.pm.

Coase, Ronald. "The Problem of Social Cost." *Journal of Law and Economics* 3 (October 1960): 1–44.

Coffin, James B. "Clinton Administration Again Recommends Permanent Fee Demo." *Federal Parks & Recreation,* February 25, 2000, 8.

Cooke, Alistair. *Alistair Cooke's America.* New York: Knopf, 1973.

Copes, Parzival. "A Critical Review of the Individual Quota as a Device in Fisheries Management." *Land Economics* 62 (August 1986): 278–91.

Crandall, Robert W. "Ackerman and Hassler's *Clean Coal/Dirty Air.*" *Bell Journal of Economics* 1 (autumn 1981): 678.

Crawford, Mark. "Scientists Battle Over Grand Canyon Pollution." *Science* 247 (February 23, 1990): 911–12.

Crutchfield, J. A., and G. Pontecorvo. *The Pacific Salmon Fisheries: A Study of Irrational Conservation*. Baltimore: Johns Hopkins University Press, for Resources for the Future, 1969.

Crutchfield, James A. "Resources from the Sea." In *Ocean Resources and Public Policy,* edited by T. S. English. Seattle: University of Washington Press, 1973.

Current, Richard N. *Pine Logs and Politics: A Life of Philetus Sawyer, 1816–1900*. Madison: State Historical Society of Wisconsin, 1950.

Dallas Times Herald, July 16, 1984, B3.

Daly, George, and Thomas Mayor. "Equity, Efficiency and Environmental Quality." *Public Choice* 51 (1986): 154.

Daly, Herman E. *Steady-State Economics.* San Francisco: W. H. Freeman and Company, 1977.

Davignon, Jean-Francois. "The Quebec Story." Quebec, Canada: Atlantic Salmon Federation, n.d.

De Alessi, Michael. *Fishing for Solutions.* London, England: The Institute of Economic Affairs, 1998.

———. "Fishing for Solutions: The State of the World's Fisheries." In *Earth Report 2000: Revisiting the True State of the Planet,* edited by Ronald Bailey. New York: McGraw-Hill, 2000.

De Alessi, Louis. "Private Property Rights as the Basis for Free Market Environmentalism." In *Who Owns the Environment?* edited by Peter J. Hill and Roger E. Meiners. Lanham, MD: Rowman and Littlefield Publishers, 1998.

De Laguna, Frederica. *The Story of a Tlingit Community.* Bureau of American Ethnology Bulletin 172. Washington, DC: U.S. Government Printing Office, 1972.

Deacon, Robert T. "Deforestation and the Rule of Law in a Cross Section of Countries." *Land Economics* 70 (1994): 414–30.

Defenders of Wildlife. *1997 Annual Report*. Washington, DC: Defenders of Wildlife, 1997.

Defenders of Wildlife. "Wolf Compensation Trust." May 12, 2000. Available: http://www.defenders.org/wolfcomp.html.

Delta Waterfowl. *Delta Waterfowl: Adopt a Pothole Summary Report*. Deerfield, Illinois, August 1993.

———. *Delta Waterfowl Report*. Deerfield, Illinois, fall 1994.

Demsetz, Harold. "Toward a Theory of Property Rights." *American Economic Review* 57 (May 1967): 347–59.

DeSimone, Livio D., and Frank Popoff. *Eco-Efficiency: The Business Link to Sustainable Development*. Cambridge, MA: MIT Press, 1997.

Diamond, Jared. "Paradise and Oil: In a New Guinea Rain Forest, Environmentalists and Business Executives Learn That What is Good for This Pristine World is Also Good for the Bottom Line." *Discover,* March 1999, 96.

Dodd, Gareth W. "EPA Report Says U.S. Government Nation's No. 1 Polluter of Waterways." *U.S. Water News* (May 2000): 12.

Dodge, Sue E., ed. *The Nature Conservancy Magazine* 40 (March–April 1990).

Dombeck, Michael. Forest Service chief. Congressional Testimony. March 18, 1997.

Ducks Unlimited. *1998 Annual Report*. Memphis, TN: Ducks Unlimited, 1998.

Eaton, Peter. "Customary Land Tenure and Conservation in Papua New Guinea." In *Culture and Conservation: The Human Dimension in Environmental Planning,* edited by Jeffrey A. McNeely and David Pitt. Dover, NH: Croom Helm, 1985.

Eckert, Ross. *The Enclosure of Ocean Resources.* Stanford, CA: Hoover Institution Press, 1979.

Economic Report of the President—February 1988. Washington, DC: U.S. Government Printing Office, 1988.

Efron, Edith. *The Apocalyptics.* New York: Simon and Schuster, 1984.

Ehrlich, Paul. *The Population Bomb.* New York: Sierra Club-Ballantine Book, 1968.

Ellis, A. G. "Upper Wisconsin Country." In *Collections of the State Historical Society of Wisconsin,* edited by Lyman C. Draper. Vol. 3. Madison: State Historical Society of Wisconsin, 1857.

Emel, Jacque L. "Groundwater Rights: Definition and Transfer." *Natural Resources Journal* 27 (summer 1987): 653–73.

Environmental Consultants (Pvt) Ltd. *People, Wildlife and Natural Resources—The CAMPFIRE Approach to Rural Development in Zimbabwe.* Harare, Zimbabwe: Zimbabwe Trust, 1990.

Environmental Defense Fund. *1997 Annual Report.* New York: Environmental Defense Fund, 1997.

Epstein, Kyra. "Water Trusts Provide Market-Driven Solution For Conservation, Water Quality." *U.S. Water News* 16 (May 1999): 21.

Exxon Valdez Oil Spill Trustee Council. "The Settlement." *Legacy of an Oil Spill: Ten Years After Exxon Valdez.* March 27, 2000. Available: http://www.oilspill.state.ak.us/setlement/setlment.htm

Fay Fly Fishing Properties, Inc. Published brochure, n.d. Bozeman, Montana.

Fernow, Bernhard E. *Economics of Forestry.* New York: Thomas Y. Crowell, 1902.

Field, Melissa, Schiedermayer & Associates. Letter and fact sheets to author, n.d.

Fischer, Hank. *Wolf Wars: The Remarkable Story of the Restoration of Wolves in Yellowstone.* Helena, MT: Falcon Publishing, 1995.

Fisher, Anthony. *Resource and Environmental Economics.* New York: Cambridge University Press, 1981.

"The Flaw in the Fisheries Bill." *Washington Post,* April 13, 1976.

Food and Agriculture Organization of the United Nations (FAO). *Marine Fisheries and the Law of the Sea: A Decade of Change.* Rome, Italy: FAO, 1993.

———. *The State of World Fisheries and Aquaculture 1998.* Available: http://www.fao.org/docrep/w9900e/w9900e00.htm.

———. "Water to Feed the World: Perspectives for the Future." News and Highlights. March 22, 2000. Available: http://www.fao.org/news/2000/000306–e.htm.

Francis, George. "Great Lakes Governance and the Ecosystem Approach: Where Next?" *Alternatives* 3 (September–October 1986): 66.

Fretwell, Holly Lippke. *Public Lands: The Price We Pay.* Public Lands Report No. 1. Bozeman, MT: Political Economy Research Center, August 1998.

———. *Forests: Do We Get What We Pay For?* Public Lands Report No. 2. Bozeman, MT: Political Economy Research Center, July 1999.

———. "Paying To Play: The Fee Demonstration Program." *PERC Policy Series* No. PS-17. Bozeman, MT: Political Economy Research Center, December 1999.

———. *Federal Estate: Is Bigger Better?* Public Lands Report No. 3. Bozeman, MT: Political Economy Research Center, May 2000.

Friedman, Milton, and Anna J. Schwartz. *A Monetary History of the United States.* Princeton, NJ: Princeton University Press, 1963.

Fries, Robert F. *Empire in Pine: The Story of Lumber in Wisconsin.* Madison: State Historical Society of Wisconsin, 1951.

Frink, Maurice, W. Turrentine Jackson, and Agnes Wright Spring. *When Grass Was King.* Boulder: University of Colorado Press, 1956.

"FWP Commission Rejects Ruby River Closure Request." Helena AP. *Bozeman Daily Chronicle,* February 5, 1999.

Gardner, Delworth B. *Plowing Ground in Washington: The Political Economy of U.S. Agriculture.* San Francisco: Pacific Research Institute for Public Policy, 1995.

Gates, Paul W. *The Wisconsin Pine Lands of Cornell University: A Study in Land Policy and Absentee Ownership.* 2d ed. Madison: State Historical Society of Wisconsin, 1965.

———. *History of Public Land Law Development.* Washington, DC: Public Land Law Review Commission, 1968: 534–55.

Getz, Wayne M., Louise Fortmann, David Cumming, Johan du Toit, Jodi Hilty, Rowan Martin, Michael Murphree, Norman Owen-Smith, Anthony M. Starfield, Michael I. Westphal. "Sustaining Natural and Human Capital: Villagers and Scientists." *Science,* March 19, 1999.

Glick, Dennis, David Cowan, Robert Bonnie, David Wilcove, Chris Williams, Dominick Dellasala, and Steve Primm. *Incentives for Conserving Open Lands in Greater Yellowstone.* Bozeman, MT: Greater Yellowstone, 1998.

Global 2000 Report to the President. Washington, DC: U.S. Government Printing Office, 1980.

Goerold, W. Thomas. "Environmental and Petroleum Resource Conflicts: A Simulation Model to Determine the Benefits of Petroleum Production in the Arctic National Wildlife Refuge, Alaska." *Materials and Society* 11 (1987): 279–307.

Goklany, Indur. "Richer is Cleaner." In *The True State of the Planet,* edited by Ronald Bailey. New York: The Free Press, 1995.

———. *Clearing the Air.* Washington, DC: Cato Institute, 1999.

———. "Empirical Evidence Regarding the Role of Naturalization in Improving U.S. Air Quality." In *The Common Law and the Environment,* edited by Roger E. Meiners and Andrew P. Morriss. Lanham, MD: Rowman and Littlefield Publishers, 2000.

Goldberg, Carey. "Fish Farms Breed Fight Over Way of Life." *New York Times* on the web, http://www.nytimes.com, August 28, 1999.

Goldsmith, W., and T. H. Haas. "Possessory Rights of the Natives of Southeastern Alaska." Report to the Commissioner of Indian Affairs, Bureau of Indian Affairs, U.S. Department of the Interior, Juneau, Alaska, 1946.

Gordon, H. Scott, "The Economic Theory of a Common Property Resource: The Fishery." *Journal of Political Economy* 62 (April 1954): 124–42.

Gorte, Ross W. "Wilderness: Overview and Statistics." Congressional Research Service Report for Congress. December 2, 1994. Available: http://www.cnie.org/nle/nrgen_5.html.

Graff, Thomas J. "Future Water Plans Need a Trickle-Up Economizing." *Los Angeles Times,* June 14, 1982, V-2.

Grafton, R. Quentin. "Performance of and Prospects for Rights-Based Fisheries Management in Atlantic Canada." In *Taking Ownership: Property Rights and Fishery Management on the Atlantic Coast,* edited by Brian Lee Crowley. Halifax, Nova Scotia, Canada: Atlantic Institute for Market Studies, 1996.

"Green Economics." *The Economist,* June 24, 1989, 48.

Greenpeace USA. *1997 Annual Report.* Washington, DC: Greenpeace USA, 1997.

Greve, Frank. "Senior Legislators Claim Funds For Pet Projects." *Washington Post,* December 1, 1997.

Guldin, Richard W. *An Analysis of the Water Situation in the United States: 1989–2040.* USDA Forest Service General Technical Report RM-177. Washington, DC: U.S. Government Printing Office, 1989.

Gwartney, James D., and Richard L. Stroup. *Economics: Private and Public Choice.* 8th ed. Orlando, FL: Harcourt Brace & Company, 1997.

Gwartney, James, Robert Lawson, and Walter Block. *Economic Freedom of the World, 1975–1995.* Vancouver, BC, Canada: The Fraser Institute, 1996.

Hardin, Garrett. "The Tragedy of the Commons." *Science* 162 (December 1968): 1243–48.

Hartley, Peter. *Conservation Strategies for New Zealand.* Wellington, New Zealand: New Zealand Business Roundtable, 1997.

Hartwick, John M., and Nancy D. Olewiler. *The Economics of Natural Resource Use.* New York: Harper & Row, 1986.

Hayek, F. A. "The Use of Knowledge in Society." *The American Economic Review* 35 (September 1945): 519–30.

Hays, Samuel P. *Conservation and the Gospel of Efficiency: The Progressive Conservation Movement, 1890–1920.* Cambridge, MA: Harvard University Press, 1959.

Herr, Andrew. "Saving the Snake River Salmon: Are We Solving the Right Problem?" *PERC Working Paper,* no. 94-15. Bozeman, MT: Political Economy Research Center, August 26, 1994.

Herrington, William C. "Operation of the Japanese Management System." In *Alaska Fisheries Policy,* edited by Arlon R. Tussing, Thomas A. Morehouse, and James D. Babb, Jr. Fairbanks: Institute of Social, Economic and Government Research, 1972.

Hide, Rodney P., and Peter Ackroyd. "Depoliticising Fisheries Management: Chatham Islands' Paua (Abalone) as a Case Study." Working paper. Christchurch, New Zealand: Centre for Resource Management, Lincoln University, March 1990.

Higgins, Margot. "Atlantic Salmon Protection Expected." *Environmental News Network,* October 15, 1999. Available: http://www.enn.com/enn-news-archive/1999/10/101599/asalmon_6485.asp.

Higgs, Robert. "Legally Induced Technical Regress in the Washington Salmon Fishery." *Research in Economic History* 7 (1982): 55–86.

Holcombe, Randall G. "Urban Sprawl: Pro and Con." *PERC Reports* 17(1). Bozeman, MT: Political Economy Research Center, February 1999: 3–5.

Hotelling, Howard. "The Economics of Exhaustible Resources." *Journal of Political Economy* 39 (1931): 137–75.

"How Brazil Subsidises the Destruction of the Amazon," *The Economist,* March 18, 1989, 69.

Huffman, James. "Instream Uses: Public and Private Alternatives." In *Water Rights: Scarce Resource Allocation, Bureaucracy, and the Environment,* edited by Terry L. Anderson. San Francisco: Pacific Institute for Public Policy Research, 1983.

Hurt III, Harry, Lynda Wright, and Pamela Abramson. "Alaska After." *Newsweek,* September 18, 1989.

Hutchins, Wells A. *Water Rights Laws in the Nineteen Western States.* Miscellaneous Publication no. 1206. Vol. 1. Washington, DC: Natural Resources Economics Division, U.S. Department of Agriculture, 1971: 442–54.

Hyde, Dayton O. "Recreation and Wildlife on Private Lands." In *Recreation on Private Lands: Issues and Opportunities,* by the Task Force on Recreation on Private Lands. Workshop proceedings. Washington, DC, March 10, 1986.

Hylton, Victoria. "The Wild Harvest." Johannesburg, South Africa: Southern Wild Productions, n.d.

Idaho Department of Fish and Game (IDFG). "Can We Bring Back Salmon Fishing?" In *Saving Idaho's Salmon.* Published pamphlet. IDFG, Boise, Idaho.

Interagency Wild and Scenic Rivers Coordinating Council and the National Trails System Map and Guide-Text. Available: http://www.nps.gov/htdocs1/pub_aff/naltrail.htm. Accessed: February 11, 1998.

Iversen, Edwin S., and Jane Z. Iversen. "Salmon-farming Success in Norway." *Sea Frontiers* (November–October 1987): 355–61.

Ives, Angela. "Love Canal." In *The Market Meets the Environment,* edited by Bruce Yandle. Lanham, MD: Rowman and Littlefield Publishers, 1999.

Izaak Walton League of America. *1997 Annual Report.* Gaithersburg, MD: Izaak Walton League of America, 1997.

Jablonski, Karol. Delta Waterfowl Foundation. Telefax to author. May 15, 2000.

Jackson, W. Turrentine. "The Wyoming Stock Growers Association, Its Years of Temporary Decline, 1886–1890." *Agricultural History* 22 (October 1948).

Jentoft, Svein. "Fisheries Co-management: Delegating Responsibility to Fishermen's Organizations." *Marine Policy* (April 1989): 137–54.

Jentoft, Svein, and Trond Kristoffersen. "Fishermen's Co-management: The Case of the Lofoten Fishery." *Human Organization* 48 (1989): 355–65.

John Cordell, ed. *A Sea of Small Boats.* Cambridge, MA: Cultural Survival, 1989.

Johnsen, D. Bruce. "The Formation and Protection of Property Rights among Southern Kwakiutl Indians." *Journal of Legal Studies* 15 (January 1986): 41–67.

Johnson, R. W. "Public Trust Protection for Stream Flows and Lake Levels." *University of California at Davis Law Review* 14 (1980): 256–57.

Johnson, Ronald N. "Implications of Taxing Quota Value in an Individual Transferable Quota Fishery." *Marine Resource Economics* 10 (1995): 327–40.

Johnson, Ronald N., and Gary D. Libecap. *Explorations in Economic History* 17 (1980): 376–77.

———. "Contracting Problems and Regulation: The Case of the Fishery." *American Economic Review* 12 (December 1982): 1005–22.

Kane, Lucile. "Federal Protection of Public Timber in the Upper Great Lakes States." *Agricultural History* 23 (1949): 135–39.

Karpiak, Steven. "The Establishment of Porcupine Mountains State Park." *Michigan Academician* 2 (1978): 135–39.

Keiter, Robert B. "Preserving Nepal's National Parks: Law and Conservation in the Developing World." *Ecology Law Quarterly* 22 (1995): 637.

Keynes, John Maynard. *The General Theory of Employment, Interest, and Money.* New York: Harcourt, Brace & World, 1964.

Kinney, Clesson S. *Law of Irrigation and Water Rights and the Arid Region Doctrine of Appropriation of Waters.* Vol. 1. San Francisco: Bender-Moss, 1912, sec. 598.

Knight, Jr., James Everett. "Effect of Hydrocarbon Development on Elk Movements and Distribution in Northern Michigan." Ph.D. diss., University of Michigan, Ann Arbor, 1980.

Kreuter, Urs P., and Randy T. Simmons. "Economics, Politics and Controversy Over African Elephant Conservation." In *Elephants and Whales: Resources for Whom?* edited by Milton M. R. Freeman and Urs P. Kreuter. Switzerland: Gordon and Breach Science Publishers, 1994.

Lamm, Richard D. "Forward." In *Western Water: Tuning the System,* by Bruce Driver. Denver: Western Governors' Association, 1986.

"Land Trusts Are Booming." *Bozeman Daily Chronicle,* December 4, 1998.

Landry, Clay J. "Market Transfers of Water for Environmental Protection of the Western United States." Working Paper WP98-3. Political Economy Research Center, Bozeman, Montana, 1998.

———. *Saving Our Streams Through Water Markets: A Practical Guide.* Bozeman, MT: Political Economy Research Center, 1998.

Langdon, Steve. "From Communal Property to Common Property of Limited Entry: Historical Ironies in the Management of Southeast Alaska Salmon." In *A Sea of Small Boats,* edited by John Cordell. Cambridge, MA: Cultural Survival, 1989.

Larson, Agnes M. *History of the White Pine Industry in Minnesota.* Minneapolis: University of Minnesota Press, 1949.

Leal, Donald R. "Unlocking the Logjam Over Jobs and Endangered Animals." *San Diego Union-Tribune,* April 18, 1993.

———. "Turning a Profit on Public Forests." *PERC Policy Series* No. PS-4. Bozeman, MT: Political Economy Research Center, September 1995.

———. "Community-Run Fisheries: Avoiding the Tragedy of the Commons." *PERC Policy Series* No. PS-7. Bozeman, MT: Political Economy Research Center, September 1996.

Leal, Donald R., and Holly Lippke Fretwell. "Back to the Future to Save Our Parks." *PERC Policy Series* No. PS-10. Bozeman, MT: Political Economy Research Center, June 1997.

———. "Parks in Transition: A Look at State Parks." Bozeman, MT: Political Economy Research Center, June 1997. Available: http://www/perc.org/stpk.htm.

Leal, Donald R., and J. Bishop Grewell. *Hunting for Habitat: A Practical Guide to State-Landowner Partnerships.* Bozeman, MT: Political Economy Research Center, 1999.

Leffler, Merrill. "Killing 'Maryland's Oysters." *Washington Post,* March 29, 1987.

Leopold, Aldo. "Conservation Economics." In *The River of the Mother of the Mother of God and Other Essays by Aldo Leopold,* edited by Susan L. Flader and J. Baird Callicott. Madison: University of Wisconsin Press, 1991 [1934].

———. "Game and Wild Life Conservation [1932]." In *The River of the Mother of God and Other Essays by Aldo Leopold,* edited by Susan L. Flader and J. Baird Callicott. Madison: University of Wisconsin Press, 1991.

Leopold, Aldo. *Report on Huron Mountain Club.* Huron, MI: Huron Mountain Club, 1938.

Libecap, Gary D. *Locking Up the Range: Federal Land Control and Grazing.* San Francisco: Pacific Institute for Public Policy Research, 1981.

Lueck, Dean. "The Law and Politics of Federal Wildlife Preservation." In *Political Environmentalism: Going Behind the Green Curtain,* edited by Terry L. Anderson. Stanford, CA: Hoover Institution Press, 2000.

Lueck, Dean. "The Economic Organization of Wildlife Institutions." In *Wildlife in the Marketplace,* edited by Terry L. Anderson and Peter J. Hill. Lanham, MD: Rowman and Littlefield Publishers, 1995.

Lueck, Dean, and Jeffrey Michael. "Preemptive Habitat Destruction Under the Endangered Species Act." Unpublished manuscript. Department of Agricultural Economics and Economics, Montana State University, Bozeman, Montana, 1999.

Maggio, Gregory F. "Recognizing the Vital Role of Local Communities in International Legal Instruments for Conserving Biodiversity." *UCLA Journal of Environmental Law and Policy* 16 (1997/1998): 179–226.

Maranto, Gina. "Caught in Conflict." *Sea Frontiers* 35 (May–June 1988): 144–51.

Margat, Jean. "A Hidden Asset." *UNESCO Courier* 15 (1993): 15.

Martin, Richard. "Resisting an Oil Rig Invasion." *Insight,* March 14, 1988, 17–18.

Marzulla, Nancy G., and Roger J. Marzulla. *Property Rights: Understanding Government Takings and Environmental Regulation.* Rockville, MD: Government Institute, 1997.

Mattison, Ray H. "The Hard Winter and the Range Cattle Business." *The Montana Magazine of History* 1 (October 1951): 18.

McAllister, Sean T. "Community-Based Conservation: Restructuring Institutions to Involve Local Communities in a Meaningful Way." *Colorado Journal of International Environmental Law and Policy* 10 (winter 1999): 195–225.

McClurg, Tom. "Bureaucratic Management versus Private Property: ITQs in New Zealand after Ten Years." In *Fish Or Cut Bait,* edited by Laura Jones and Michael Walker. Vancouver, BC: The Fraser Institute, 1997.

McHugh, J. L. "Jeffersonian Democracy and the Fisheries." In *World Fisheries Policy: Multidisciplinary Views,* edited by B. J. Rothschild. Seattle: University of Washington Press, 1972.

McKean, Margaret A. "Management of Traditional Common Lands (*Iriachi*) in Japan." In *Proceedings of the Conference on Common Property Resource Management, April 21–26, 1985.* National Research Council. Washington, DC: National Academy Press, 1986.

McKee, Russell. "Tombstones of a Lost Forest." *Audubon* 90 (March 1988): 68.

McMillion, Scott. "Gallatin Forest Is A Gift To Community That Costs Taxpayers." *Bozeman Daily Chronicle,* August 31, 1997.

Mead, Walter J., and Philip Sorenson. "The Economic Cost of Santa Barbara Oil Spill." In *Santa Barbara Oil Spill: An Environmental Inquiry.* Santa Barbara: California Marine Science Institute, University of California at Santa Barbara, 1972.

Meadows, Donnella H., Dennis L. Meadows, Jorgen Randers, William W. Behrens III. *The Limits to Growth: A Report for the Club of Rome's Project on the Predicament of Mankind.* New York: A Potomac Associates Book, New American Library, 1974.

Meiners, Roger E., and Bruce Yandle. "The Common Law: How It Protects the Environment." *PERC Policy Series* No. PS-13. Bozeman, MT: Political Economy Research Center, May 1998.

———. "Common Law and the Conceit of Modern Environmental Policy." *George Mason Law Review* 7 (summer 1999): 923–82.

Menell, Peter S. "Institutional Fantasylands: From Scientific Management to Free Market Environmentalism." *Harvard Journal of Law and Public Policy* 15 (1992): 589–610.

Merk, Frederick. *Economic History of Wisconsin During the Civil War Decade.* Madison: State Historical Society of Wisconsin, 1916.

"The Mess One Man Makes." *The Economist,* April 22, 2000, 19–22.

Messerschmidt, D. A. "Collective Management of Forest Hills in Nepal." In *Proceedings of the Conference on Common Property Resource Management.* National Research Council. Washington, DC: National Academy Press, 1986.

Michalak, Anna M. "Feasibility of Contaminant Source Identification for Property Rights Enforcement." Paper presented at "The Technology of Property Rights," 1999 PERC Political Economy Forum, Bozeman, Montana, December 2–5, 1999.

Milazzo, Mateo. "Subsidies in World Fisheries: A Reexamination." World Bank Technical Paper No. 406 (Fisheries Series). Washington, DC: National Academy Press, 1999.

Miller, H. Crane. *Turning the Tide on Wasted Tax Dollars: Potential Federal Savings from Additions to the Coastal Barrier Resources System.* Washington, DC: National Wildlife Federation, April 17, 1989.

Milstein, Michael. "Park Water Risky." *Billings* [Montana] *Gazette,* July 5, 1998.

———. "GAO Report Praises Park Fee System." *Billings* [Montana] *Gazette,* February 16, 1999.

Mitchell, John G. "The Oil Below." *Audubon* 83 (May 1981): 16–17.

———. "Our National Parks: Legacy at Risk." *National Geographic* 186 (October 1994): 54.

Mitchell, Ronald. "Intentional Oil Pollution of the Oceans." In *Institutions for the Earth: Sources of Effective International Protection,* edited by Peter M. Haas et al. Cambridge, MA: MIT Press, 1993.

Monaghan, Jay, ed. *The Book of the American West.* New York: Bonanza, 1963.

Montana Wildlife Federation. "Public Ownership of Wildlife & the Threat of Privatization." Helena, Montana. February 22, 1999.

Moore, Thomas Gale. *Climate of Fear.* Washington, DC: Cato Institute, 1998.

Morris, Julian. "International Environmental Agreements: Developing Another Path." In *The Greening of U.S. Foreign Policy,* edited by Terry L. Anderson and Henry I. Miller. Stanford, CA: Hoover Institution Press, 2000.

Morrisey, Wayne A., and John R. Justus. "IB89005: Global Climate Change." CRS Issue Brief for Congress, Congressional Research Service. March 13, 2000. Available: http://www.cnie.org/nle/clm-2.html.

Nasser, Haya El. "Gore Veers From Past Growth Policy." *USA Today,* January 12, 1999.

National Audubon Society. *1998 Annual Report.* Available: http://www.audubon.org/nas/ar/accounting.html. Accessed: March 8, 1999.

National Institute for Urban Wildlife. *Environmental Conservation and the Petroleum Industry.* Washington, DC: American Petroleum Institute, n.d.

National Marine Fisheries Service (NMFS), *The Economic Status of U.S. Fisheries, 1996,* U.S. Department of Commerce, National Oceanic and Atmospheric Administration, National Marine Fisheries Service, December 1996. Available: http://www.st.nmfs.gov/econ/oleo/oleo.html.

———. *Our Living Oceans: Report on the Status of U.S. Living Marine Resources, 1999.* U.S. Department of Commerce. NOAA Technical Memorandum NMFS-F/SPO-41. Available: http://spo.nwr.noaa.gov/fa3.pdf.

———. *Our Living Oceans: Report on the Status of U.S. Living Marine Resources, 1999.* U.S. Department of Commerce. NOAA Technical Memorandum NMFS-F/SPO-41. Available: http://spo.nwr.noaa.gov/olo99.htm.

National Marine Fisheries Service (NMFS). *Report to Congress: Status of Fisheries of the United States.* September 1998. Available: http://www.nmfs.gov/sfa/98stat.pdf.

National Parks and Conservation Association. *1998 Annual Report*. Washington, DC: National Parks and Conservation Association, 1998.

National Petroleum Council. *U.S. Arctic Oil and Gas.* Washington, DC: National Petroleum Council, 1981.

National Wilderness Preservation System home page. Available: http://www.wilderness.net/nwps/.

National Wildlife Federation. *1998 Annual Report.* Vienna, VA: National Wildlife Federation, 1998.

Natural Resources Defense Council. *1997 Annual Report.* New York: Natural Resources Defense Council, 1997.

Netting, Robert. *Balancing on an Alp.* New York: Cambridge University Press, 1981.

North, Douglass C., Terry L. Anderson, and Peter J. Hill. *Growth and Welfare in the American Past: A New Economic History.* Englewood Cliffs, NJ: Prentice-Hall, 1983, 111–21.

North Maine Woods, Inc. *North Maine Woods.* Published brochure. 1998. Available from North Maine Woods, Inc., P.O. Box 421, Ashland, Maine 04732.

Northwest Power Planning Council (NPPC). *Columbia River Basin Fish and Wildlife Program* (Typescript Copy). Portland, Oregon, February 1994.

Norton, Seth W. "Property Rights, the Environment, and Economic Well-Being." In *Who Owns the Environment?* edited by Peter J. Hill and Roger E. Meiners. Lanham, MD: Rowman and Littlefield Publishers, 1998.

Oberg, Kalervo. *The Social Economy of the Tlingit Indians.* American Ethnological Society Monograph 55. Seattle: University of Washington Press, 1973.

Office of Management and Budget. *Analytical Perspectives, Budget of the United States Government, Fiscal Year 2000.* Washington, DC, 2000.

Ogden, Bill. Telephone conversation, April 21, 2000.

Olson, R. L. "Social Structure and Social Life of the Tlingit in Alaska." *Anthropological Records* 26. Berkeley: University of California Press, 1967.

Olson, Sherry H. *The Depletion Myth: A History of Railroad Use of Timber.* Cambridge, MA: Harvard University Press, 1971.

"Once It's Here . . ." *The Economist,* March 1, 1997, 19–20.

Opaluch, James, and Richard M. Kashmanian. "Assessing the Viability of Marketable Permit Systems: An Application in Hazardous Waste Management." *Land Economics* 61 (August 1985): 263–71.

Ophuls, William. "Leviathan or Oblivion?" In *Toward a Steady-State Economy,* edited by Herman Daley. San Francisco: W. H. Freeman, 1973.

Osgood, Ernest Staples. *The Day of the Cattleman*. Minneapolis: University of Minnesota Press, 1929.

Ostrom, Elinor. *Governing the Commons: The Evolution of Institutions for Collective Action*. New York: Cambridge University Press, 1990.

Ostrom, Elinor, James Walker, and Roy Gardner. "Covenants With and Without a Sword: Self-Governance Is Possible." In *The Political Economy of Customs and Culture: Informal Solutions to the Commons Problem,* edited by Terry L. Anderson and Randy T. Simmons. Lanham, MD: Rowman and Littlefield Publishers, 1993.

O'Riordan, Timothy J. "The Politics of Sustainability." In *Sustainable Environmental Management,* edited by R. K. Rutner. Boulder, CO.: Westview Press, 1988.

O'Toole, Randal. "Learning the Lessons of the 1980s." *Forest Watch* 10 (January–February 1990): 6.

Pautzke, Clarence G., and Chris W. Oliver. *Development of Individual Fishing Quota Program for Sablefish and Halibut Longline Fisheries off Alaska.* Anchorage: North Pacific Fishery Management Council, 1977.

Pearse, Peter H., and James R. Wilson. "Local Co-management of Fish and Wildlife: The Quebec Experience." *Wildlife Society Bulletin* 27 (1999): 678.

Pigou, A. C. *The Economics of Welfare.* London, England: Macmillan, 1920.

Pinchot, Gifford. *The Fight for Conservation.* New York: Doubleday and Page, 1910.

Platts, Linda. "Environmentalists Use Market Tools." *Montana Farmer-Stockman.* February 1997, 18.

Political Economy Research Center. *Commentary by Northwest Ecosystem Alliance.* July 7, 1999. Available: http://www.perc.org/newsloom.htm.

Pope, Dennis. Biological Team Leader, GSE. E-mail to author. February 3, 1999.

Porter, Ed, and Lee Huskey. "The Regional Economic Effect of Federal OCS Leasing: The Case of Alaska." *Land Economics* 57 (November 1981): 594.

Porter, Gareth. "Too Much Fishing Fleet, Too Few Fish: A Proposal for Eliminating Global Fishing Overcapacity." Prepublication draft. World Wildlife Fund, August 1998.

Portfolio Management Task Force, World Bank. *Effective Implementation: Key to Development Impact.* Published report. Washington, DC: World Bank, October 2, 1992.

Postel, Sandra. "Facing Water Scarcity." In *State of the World 1993.* Project directed by Lester R. Brown. New York: W. W. Norton & Company, 1993.

Pound, Edward T. "Costly Outhouses Monuments to Red Tape." *USA Today,* December 15, 1997.

President's Commission on American Outdoors. *American Outdoors: The Legacy, The Challenge.* Washington, DC: Island Press, 1987.

President's Council on Environmental Quality. *15th Annual Report of the Council on Environmental Quality.* Washington, DC: U.S. Government Printing Office, 1984.

Presidio Trust. *The Presidio Trust Financial Management Program.* Report to Congress. San Francisco, California, July 8, 1998.

Price Waterhouse. *Human Resources Development Canada: Operational Review of the Atlantic Groundfish Strategy.* Ottawa, Canada: Price Waterhouse, 1995.

"Private Clubs Provide Choice Shooting." *Fishing and Hunting News* 12 (April 1982): 2.

Pub. L. No. 94-265, sec. 303(b)(6), 94th Congress, H.R. 200, April 13, 1976.

Randall, Alan. *Resource Economics: An Economic Approach to Natural Resource and Environmental Policy.* New York: Wiley, 1987.

Repetto, Robert. *The Forest for the Trees? Government Policies and the Misuse of Forest Resources.* Washington, DC: World Resources Institute, 1988.

Rich, Bruce. *Mortgaging the Earth: The World Bank, Environmental Impoverishment, and the Crisis of Development.* Boston: Beacon Press, 1994.

Richards, Bill. "Amoco Ordered to Pay Award of $85.2 Million." *Wall Street Journal,* January 12, 1988.

Ridenour, James M. *The National Parks Compromised: Pork Barrel Politics and America's Treasures.* Merrillville, IN: ICS Books, 1994.

Riggs, David W. "Market Incentives for Water Quality." In *The Market Meets the Environment,* edited by Bruce Yandle. Lanham, MD: Rowman and Littlefield Publishers, 1999.

Robbins, Jim. "Ranchers Finding Profit in Wildlife." *New York Times,* December 13, 1987.

Robinson, Jerome B. "The Next Step for Atlantic Salmon." *Field & Stream,* September 1994, 22–25.

Robinson, William L. "Individual Transferable Quotas in the Australian Southern Bluefin Tuna Fishery." In *Fishery Access Control Programs Worldwide: Proceedings of the Workshop on Management Options for the North Pacific Longline Fishers.* Alaska Sea Grant Report no. 86–4. Orca Island, WA: University of Alaska, 1986.

Rodgers III, Andrew D. *Bernhard Edward Fernow: A Story of North American Forestry.* Princeton, NJ: Princeton University Press, 1951.

Rosholt, Malcolm. *The Wisconsin Logging Book.* Rosholt, WI: Rosholt House, 1980.

Rothbard, Murray. "Law, Property Rights, and Air Pollution." *Cato Journal* 2 (spring 1982): 90.

Routhier, Yannick. Ministry of the Environment and Wildlife, Quebec, Canada. Personal communication. September 14, 1995, and November 22, 1994.

Ruddle, Kenneth, and Tomoya Akimichi. "Sea Tenure in Japan and the Southwestern Ryukus." In *A Sea of Small Boats,* edited by John Cordell. Cambridge, MA: Cultural Survival, 1989.

Rude, Kathleen. "Ponded Poisons." *Ducks Unlimited* 54 (January–February 1990): 14–18.

Runte, Alfred. *Trains of Discovery.* Niwot, CO: Roberts Rinehart, 1990.

Rydholm, Fred. "Upper Crust Camps." In *A Most Superior Land: Life in the Upper Peninsula of Michigan.* Lansing: Michigan Natural Resources Magazine, 1983.

Sagoff, Mark. "Free Market Versus Libertarian Environmentalism." *Critical Review* 6 (spring/summer 1992): 211–30.

Sand County Foundation. Telefax to author. July 28, 1995.

Sanera, Michael, and Jane S. Shaw. *Facts, Not Fear: Teaching Children About the Environment.* Washington, DC: Regnery Publishing, 1999.

Scarlett, Lynn. "New Environmentalism." *NCPA Policy Report 201.* Dallas: National Center for Policy Analysis, January 1997.

———. "Doing More with Less: Dematerialization—Unsung Environmental Triumph." In *Earth Report 2000,* edited by Ronald Bailey. New York: McGraw-Hill, 2000.

Scarlett, Lynn, and Jane S. Shaw. "Environmental Progress: What Every Executive Should Know." *PERC Policy Series* No. PS-15. Bozeman, MT: Political Economy Research Center, April 1999.

Scarlett, Lynn, Richard McCann, Robert Anex, and Alexander Volokh. "Packaging, Recycling, and Solid Waste." Reason Public Policy Institute, *Policy Study* 223, Los Angeles, California, 1997.

Schleyer, Renato Gazmuri. "Chile's Market-Oriented Water Policy: Institutional Aspects and Achievements." In *Water Policy and Water Markets,* edited by Guy Le Moigne et al. World Bank Technical Paper 249. Washington, DC: World Bank, 1994.

Schoenbrod, David. "Why States, Not EPA, Should Set Pollution Standards." In *Environmental Federalism,* edited by Terry L. Anderson and Peter J. Hill. Lanham, MD: Rowman and Littlefield Publishers, 1997.

———. "Time for the Federal Aristocracy to Give Up Power." Center for the Study of American Business, *Policy Study* 144. Washington University, St. Louis, Missouri, February 1998.

———. "Protecting the Environment in the Spirit of the Common Law." In *The Common Law and the Environment: Rethinking the Statutory Basis for Modern Environmental Law,* edited by Roger E. Meiners and Andrew P. Morriss. Lanham, MD: Rowman and Littlefield Publishers, 2000.

Schultz, Bill. *Forestry Best Management Practices Implementation Monitoring.* Missoula, MT: Montana Department of State Lands, 1992.

Schwantes, Carlos. *Railroad Signatures across the Pacific Northwest.* Seattle: University of Washington Press, 1993.

Scott, Anthony D. "Market Solutions to Open-Access, Commercial Fisheries Problems." Paper presented at Association for Public Policy Analysis and Management 10th Annual Research Conference, Seattle, October 27–29, 1988.

———. "The ITQ as a Property Right: Where it Came From, How It Works, and Where It Is Going." In *Taking Ownership: Property Rights and Fishery Management on the Atlantic Coast,* edited by Brian Lee Crowley. Halifax, Nova Scotia: Atlantic Institute for Market Studies, 1996.

Scott, Sue. "Greenland Salmon Fishery Ends." *News Release Communiqué.* Atlantic Salmon Federation, St. Andrews, New Brunswick, Canada. August 1, 1993.

Scoville, Warren. "Did Colonial Farmers 'Waste' Our Lands?" *Southern Economic Journal* 20 (1953): 178–81.

Sharrow, Barbara. Visitor Services Team Leader, GSE. E-mail to author. January 26, 1999.

Shaw, Jane S. "Environmental Regulation: How It Evolved and Where It is Headed." *Real Estate Issues* 1 (1996): 4–9.

Sheehan, James M. "The Greening of the World Bank: A Lesson in Bureaucratic Survival." Foreign Policy Briefing No. 56. Washington, DC: Cato Institute, 2000.

Short, Kevin MacEwen. "Self-management of Fishing Rights by Japanese Cooperative Associations: A Case Study from Hokkaido." In *A Sea of Small Boats,* edited by John Cordell. Cambridge, MA: Cultural Survival, 1989.

Sierra Club and subsidiaries. *1997 Annual Report.* San Francisco: Sierra Club, 1997.

Simon, Julian. *The Ultimate Resource.* Princeton, NJ: Princeton University Press, 1981.

Simon, Julian, and Herman Kahn. *The Resourceful Earth: A Response to Global 2000.* Oxford, England: Basil Blackwell, 1984.

Smets, H. "Compensation for Exceptional Environmental Damage Caused by Industrial Activities." In *Insuring and Managing Hazardous Risks: From Seveso to Bhopal and Beyond,* edited by Paul R. Kleindorfer and Howard C. Kunreuther. Berlin, Germany: Springer-Verlag, 1987.

Smith, Jr., Fred L. "Controlling the Threat of the Global Liberal Order." Paper presented to the Mont Pelerin Society. Christchurch, New Zealand, November 1989.

Smith, Robert F. Chairman, Committee on Agriculture, U.S. House of Representatives. *Statement.* Hearing to Review the Forest Service Timber Sale Program. Photocopy, June 11, 1998.

Smith, Rodney T. *Trading Water: The Legal and Economic Framework for Water Marketing.* Claremont, CA: Claremont McKenna College, Center for Study of Law Structures, 1986.

Snyder, Pamela, and Jane S. Shaw. "PC Oil Drilling in a Wildlife Refuge." *Wall Street Journal,* September 7, 1995, A14.

Solomon, Kenneth E. "South Dakota Fee Hunting: More Headaches or More Wildlife Problems on Agricultural Lands," edited by D. L. Hallett, W. R. Edwards, and G. V. Burger. Bloomington, IN: North Central Section of the Wildlife Society, 1988.

Solow, Robert M. "The Economics of Resources or the Resources of Economics." *American Economic Review* 64 (May 1974): 1–14.

Southerland, Douglas. *The Landowner.* London, England: Anthony Bond, 1968.

Sowell, Thomas. *A Conflict of Visions.* New York: William Morrow and Company, 1987.

Spencer, Roy W. "How Do We Know the Temperature of the Earth?" In *Earth Report 2000: Revisiting the True State of the Planet,* edited by Ronald Bailey. New York: McGraw-Hill, 2000.

Sporer, Christopher. "An Intelligent Tale of Fish Management." *Fraser Forum,* December 1998, 12–13.

Stackhouse, John. "Forests Returning to the Himalayas: First Nepal's Forestry Program Failed, Then the People Took Over and Saved the Trees." *Globe and Mail,* October 22, 1998.

Stalling, David. "Public Elk, Private Lands: Should Landowners Benefit from Elk and Elk Hunting?" *Bugle,* January–February 1999, 73.

Sterba, James P. "Plight of the Pheasant Frames the Debate Over Hunting's Future." *Wall Street Journal,* February 1, 1999.

Sticker, Robert R. "Commercial Fishing and Net-pen Salmon Aquaculture: Turning Conceptual Antagonism Toward a Common Purpose." *Fisheries* 13 (July–August 1988): 9–13.

Stratmann, Thomas. "The Politics of Superfund." In *Political Environmentalism: Going Behind the Green Curtain.* Stanford, CA: Hoover Institution Press, 2000.

Striker, Don. Comptroller of Yellowstone National Park. Personal communication, March 12, 1996.

Stroup, Richard L. "The Economics of Compensating Property Owners." *Contemporary Economic Policy* 15 (1997): 55–65.

———. "Superfund: The Shortcut That Failed." In *Breaking the Environmental Policy Gridlock,* edited by Terry L. Anderson. Stanford, CA: Hoover Institution Press, 1997.

Stroup, Richard L., and John A. Baden. "Endowment Areas: A Clearing in the Policy Wilderness." *Cato Journal* 2 (winter 1982): 691–708.

Sullivan, Cheryl. "Salmon Feedlots in Northwest." *Christian Science Monitor,* July 23, 1987.

Sutton, Earl. U.S. Forest Service, Northern Region. E-mail to author, March 15, 1999.

Task Force on Atlantic Fisheries [Canada]. *Navigating Troubled Waters for the Atlantic Fisheries.* Published report. Ottawa, Canada: Ministry of Supply and Services, 1982.

Task Force on Recreation on Private Lands. *Recreation on Private Lands: Issues and Opportunities.* Workshop proceedings. Washington, DC, March 10, 1986.

The National Institute for Urban Wildlife. *Environmental Conservation and the Petroleum Industry.* Washington, DC: American Petroleum Institute, n.d.

The Nature Conservancy. *1998 Annual Report.* Arlington, VA: The Nature Conservancy, 1998.

The Wilderness Society. *1998 Annual Report.* Washington, DC: The Wilderness Society, 1998.

Thompson, Andrew McFee. "Free Market Environmentalism and the Common Law: Confusion, Nostalgia, and Inconsistency." *Emory Law Journal* 45 (fall 1996): 1329–72.

Thresher, Valerie. "Economic Reflections on Wildlife Utilization in Zimbabwe." Master's thesis, University of California at Davis, 1993.

Tiebout, Charles M. "A Pure Theory of Local Expenditures." *Journal of Political Economy* 64 (1956): 416–24.

Tietenberg, Tom. *Environmental and Natural Resource Economics.* Glenview, IL: Scott, Foresman and Company, 1984.

———. *Environmental and Natural Resource Economics.* 2d ed. Glenview, IL.: Scott, Foresman and Company, 1988.

Todd, David. "Common Resources, Private Rights and Liabilities: A Case Study on Texas Groundwater Law." *Natural Resources Journal* 32 (summer 1992): 233–63.

Trelease, Frank J. "Developments on Groundwater Law." In *Advances in Groundwater "Mining" in the Southwestern States,* edited by Z. A. Saleem. Minneapolis: American Water Resources Association, 1976.

Trout Unlimited. *1997 Annual Report.* Available: http://www.tu.org/whatis/97annrep.html. Accessed: March 8, 1999.

Tullock, Gordon. "The Welfare Costs of Tariffs, Monopolies, and Theft." *Western Economic Journal* 5 (June 1967): 224–32.

U.S. Department of Agriculture, Bureau of Agricultural Economics. *Livestock on Farms, January 1, 1867–1935.* Washington, DC: U.S. Government Printing Office, 1938.

U.S. Department of Commerce, Bureau of the Census. *Statistical Abstract of the United States, 1987.* Washington, DC, 1987.

U.S. Department of the Interior, Fish and Wildlife Service, and U.S. Department of Commerce, Bureau of the Census. *1996 National Survey of Fishing, Hunting, and Wildlife-Associated Recreation.* Washington, DC, November 1997.

U.S. Department of the Interior, Minerals Management Service. "Offshore Oil Production Accounts for Little of World's Ocean Pollution." News release. Washington, DC, July 26, 1983.

———. *Mineral Revenues 1997: Report on Receipts from Federal And Indian Leases.* Denver: Minerals Management Services, 1997.

U.S. Department of the Interior, U.S. Fish and Wildlife Service. *Draft Arctic National Wildlife Refuge, Alaska Coastal Plain Resource Assessment: Report and Recommendation to the Congress of the United States and Legislative Environmental Impact Statement.* November 1986.

U.S. Environmental Protection Agency. *Draft Framework for Watershed-Based Trading.* EPA 800-R-96–001. Washington, DC: U.S. Environmental Protection Agency, May 1996.

U.S. General Accounting Office (GAO). *Federal Lands: Information on Land Owned and on Acreage with Conservation Restrictions.* GAO/RCED-95–73FS. Washington, DC, January 1995.

———. *Superfund: Half the Sites Have All Cleanup Remedies in Place or Completed.* GAO/RCED-99–245, Washington, DC, July 1995.

———. *Forest Service Decision-Making: A Framework for Improving Performance.* GAO/RECD-97–71. Washington, DC, 1997.

———. *Superfund: Progress, Problems and Future Outlook,* GAO/T-RCED-99–128, Washington, DC, March 23, 1999.

U.S. Geological Survey (USGS). *Arctic National Wildlife Refuge, 1002 Area, Petroleum Assessment, 1998.* Assessment Results. Available: http://energy.usgs.gov/factsheets/ANWR/results.html.

"U.S. Park Service Spends $333,000 on Outhouse." *ENN Daily News,* October 8, 1997.

U.S. Senate, Committee on Commerce. *A Legislative History of the Fishery Conservation and Management Act of 1976.* Washington, DC: U.S. National Marine Fisheries Service, October 1976.

Van Voorst, Bruce. "The Recycling Bottlenecks." *Time,* September 14, 1992.

Viscusi, W. Kip, and James T. Hamilton. "Are Risk Regulators Rational? Evidence from Hazardous Waste Cleanup Decisions." Working Paper No. 99–2. AEI-Brookings Joint Center for Regulatory Studies, Washington, DC, April, 1999.

Wahl, Richard W. "Cleaning Up Kesterson." *Resources* 83 (spring 1986): 12.

———. *Markets for Federal Water: Subsidies, Property Rights, and the Bureau of Reclamation.* Washington, DC: Resources for the Future, 1989.

Warren, George F., and Frank A. Pearson. *Prices.* New York: John Wiley and Sons, 1933.

Watson, Traci. "Clinton Pushes Preservation Plan." *USA Today,* January 12, 1999.

———. "Environmental Groups Wielding Power of the Purse." *USA Today,* February 3, 2000.

Webb, Walter Prescott. *The Great Plains.* New York: Grosset & Dunlap, 1931.

Weber, Gregory S. "Twenty Years of Local Groundwater Export Legislation in California: Lessons from a Patchwork Quilt." *Natural Resources Journal* 34 (summer 1994): 657–749.

Wells, Ken. "U.S. Oil Leasing Plan Is Challenged by Eskimos Trying to Protect Their Culture at World's Edge." *Wall Street Journal,* March 12, 1986.

Wernick, Iddo, Paul Waggoner, and Jesse Ausubel. "Searching for Leverage to Conserve Forests: The Industrial Ecology of Wood Products in the United States." *Journal of Industrial Ecology* 1(3): 125–45.

Wiggins, Stephen N., and Gary D. Libecap. "Oil Field Unitization: Contractual Failure in the Presence of Imperfect Information." *American Economic Review* 75 (June 1985): 368–85.

Wildavsky, Aaron. *Searching for Safety.* New Brunswick, NJ: Transaction Books, 1988.

Williamson, Lonnie L. "Wildlife Superbill Reintroduced." *Outdoor News Bulletin,* January 29, 1999.

Wiltse, Eric. "Irrigation Spells Death for Hundreds of Ruby River Trout." *Bozeman Daily Chronicle,* May 12, 1987.

"Workers Combat 25 Mile Oil Slick in Gulf of Mexico." The Associated Press, October 3, 1998. Available: http://www.willjohnston.com/articles_98/october98/10_3_98 wctosigom.html.

Woutat, Donald. "Stakes Are High in the Battle Over Exploration in Alaska National Wildlife Refuge." *Bozeman Daily Chronicle,* November 5, 1987.

Yandle, Bruce. *Common Sense and Common Law for the Environment.* Lanham, MD: Rowman and Littlefield Publishers, 1997.

———. "Environmental Regulation: Lessons from the Past and Future Prospects." In *Breaking the Environmental Policy Gridlock,* edited by Terry L. Anderson. Stanford, CA: Hoover Institution Press, 1997.

———. "Bootleggers, Baptists, and Global Warming." *PERC Policy Series* No. PS-14. Bozeman, MT: Political Economy Research Center, November 1998.

———. "Coase, Pigou, and Environmental Rights." In *Who Owns the Environment?* edited by Peter J. Hill and Roger E. Meiners. Lanham, MD: Rowman and Littlefield Publishers, 1998.

———. "The Commons: Tragedy or Triumph?" *The Freeman,* April 1999.

"Yellowstone Closes Part of the Grand Loop Road." *Island Park News,* August 14, 1998.

Yu, Daniel. Member Service Center, The Nature Conservancy. E-mail to author, March 9, 2000.

Zern, Ed. "By Yon Bonny Banks." *Field and Stream* 86 (September 1981): 120–23.

Zuesse, Eric. "Love Canal: The Truth Seeps Out." *Reason* 12 (February 1981): 16–33.

CASES CITED

Anderson v. American Smelting & Refining Co., 265 F. 928 D. Utah (1919).

Anne Anderson et al. v. Cryovac, Inc., et al., 96 F.R.D. (1983).

Anne Anderson et al. v. Cryovac, Inc., et al., 862 F.2d 910 (1988).

Anne Anderson et al. v. Grace Co. et al., 628 F. Supp. 1219 (1986).

California Trout, Inc., v. State Water Resources Control Board, 90 Cal. App. Ed 816, 153 Cal. Rptr. 672 (1979).

Carmichael v. City of Texarkana, 94 F. 561, W.D. Ark. (1899).

Central Arizona Water Conservation District et al. v. United States Environmental Protection Agency, 990 F.2d 1531 (1993).

Colorado River Water Conservation District v. Rocky Mountain Power Company, 158 Colo. 331, 406 P. 2d 798 (1965).

Ethyl Corporation et al. v. Environmental Protection Agency, 176 U.S. App. DC 373; 541 F. 2d 1 (1976).

Fullerton v. California State Water Resources Control Board, 90 Cal. App. 3d 590, 153 Cal. Rptr. 518 (1979).

Georgia v. Tennessee Copper Co., 206 U.S. 230, 27 S.Ct. 618 (1907).

Illinois v. Milwaukee, 406 U.S. 91 (1972).

Lake Shore Duck Club v. Lake View Duck Club, 50 Utah 76, 309 (1917).

Milwaukee v. Illinois, 451 U.S. 304 (1981).

New York v. Monarch Chemicals et al., 456 N.Y.S.2d 867 (1982).

New York v. Schenectady Chemicals, Inc., 459 N.Y.S.2d 971 (1983).

Secretary of the Interior et al. v. California et al., 52 U.S.L.W. 4063 (U.S. January 11, 1984), 683 F. 2d 1253 reversed (9th Cir. 1982).

Trail Smelter case, 3 RIIA 1905 (1949), 1965.

Whalen v. Union Bag & Paper Co., 208 N.Y. 1, 101 N. E. 805 (1913).

Wood v. Picillo, 443 A.2d 1244 S.Ct., R.I. (1982).

INDEX